MW01618104

Tactile Learning Activities in Mathematics

A Recipe Book for the Undergraduate Classroom

AMS / MAA | CLASSROOM RESOURCE MATERIALS

VOL 54

Tactile Learning Activities in Mathematics

A Recipe Book for the Undergraduate Classroom

Julie Barnes
Jessica M. Libertini

Providence, Rhode Island

Classroom Resource Materials Editorial Board

Susan G. Staples, Editor

Jennifer A. Bergner
Christina Eubanks-Turner
Christoper Hallstrom
Cynthia J. Huffman
Brian Paul Katz
Haseeb A. Kazi
Paul R. Klingsberg
Brian Lins
Mary Eugenia Morley
Darryl Yong

2010 *Mathematics Subject Classification.* Primary 97-01, 00-01, 26-01, 97-00.

For additional information and updates on this book, visit
www.ams.org/bookpages/clrm-54

Pringles, M&Ms, Hershey, OREO, Skittles, Jolly Rancher, Frisbee, SET, Wikki Stix, Bendaroos, and LEGO are all registered trademarks.

Library of Congress Cataloging-in-Publication Data

Names: Barnes, Julie, 1968– author. | Libertini, Jessica M., 1977– author.
Title: Tactile learning activities in mathematics: A recipe book for the undergraduate classroom / Julie Barnes, Jessica M. Libertini.
Description: Providence, Rhode Island: MAA Press, an imprint of the American Mathematical Society, 2018. | Series: Classroom resource materials; volume 54 | Includes bibliographical references and index.
Identifiers: LCCN 2017057965 | ISBN 9781470443511 (alk. paper)
Subjects: LCSH: Mathematics–Study and teaching (Higher)–Activity programs. | Mathematical recreations. | Touch. | Educational innovations. | Teaching–Methodology. | AMS: Mathematics education – Instructional exposition (textbooks, tutorial papers, etc.). msc
Classification: LCC QA20.G35 T33 2018 | DDC 510.71/1–dc23
LC record available at https://lccn.loc.gov/2017057965

Color graphic policy. Any graphics created in color will be rendered in grayscale for the printed version unless color printing is authorized by the Publisher. In general, color graphics will appear in color in the online version.

Copying and reprinting. Individual readers of this publication, and nonprofit libraries acting for them, are permitted to make fair use of the material, such as to copy select pages for use in teaching or research. Permission is granted to quote brief passages from this publication in reviews, provided the customary acknowledgment of the source is given.

Republication, systematic copying, or multiple reproduction of any material in this publication is permitted only under license from the American Mathematical Society. Requests for permission to reuse portions of AMS publication content are handled by the Copyright Clearance Center. For more information, please visit www.ams.org/publications/pubpermissions.

Send requests for translation rights and licensed reprints to reprint-permission@ams.org.

© 2018 by the American Mathematical Society. All rights reserved.
The American Mathematical Society retains all rights
except those granted to the United States Government.
Printed in the United States of America.

♾ The paper used in this book is acid-free and falls within the guidelines
established to ensure permanence and durability.
Visit the AMS home page at https://www.ams.org/

10 9 8 7 6 5 4 3 2 1 23 22 21 20 19 18

Contents

Acknowledgments

We would like to extend our thanks to Carolyn Yackel and Brian Winkel; this book would not have been written without their suggestions and advice. In fact, the very idea of writing this book started when Carolyn approached us at our Mathematical Association of America (MAA) Contributed Paper session at the 2013 Joint Mathematics Meetings in San Diego; Carolyn not only suggested that this book be written, but she also emailed Julie a few weeks later to stress the importance of such a book. Brian, who has served as a valued role model and mentor for both of us, agreed with Carolyn, advised us on how to proceed, suggested a format for the book, and even provided us with a working retreat location at his home.

We greatly appreciate the efforts of the individuals and groups who helped proofread or test some of the activities; these people include Karen Bliss, Meagan Herald, Kim Johnson, Ben Kearns, Laura Lembeck, Lisa Lyford, John Wagaman, and the Women's Writing Group at Virginia Military Institute. We would also like to thank the Western Carolina University Math Club for agreeing to pose for photographs.

This book includes activities from talented faculty members that work at a wide range of institutions, represent all career stages, and are from a variety of geographical regions across the United States. We would like to thank all of our contributing authors for sharing their ideas in this book, and for working so cooperatively with us throughout the editing process.

We would like to thank everyone involved with the Classroom Resource Materials series of the MAA's publishing division for their encouragement and support throughout the process, including editors Jerry Bryce and Susan Staples, as well as the thoughtful team of anonymous reviewers who went through this manuscript and provided us with clear and helpful guidance that shaped the revision process.

Finally, we would like to share our deepest gratitude for our dear friends and family for their support, encouragement, and understanding as we worked on this project.

Preface

Tell me, and I forget. Teach me, and I may remember. Involve me, and I learn.

—Ancient Proverb

Suppose you are scheduled to visit a college math classroom and find feather boas, hula hoops, cards, balls, jacks, or a large stack of cookies. Perhaps you think the class is somewhere else, and this classroom is set up for young children. Thinking back on your own preschool days, you know that using tactile activities for learning is commonly accepted and well documented in early education; the learning theories of Froebel, Montessori, Piaget, and Erikson all indicate the importance of hands-on learning. While you may have heard of experiential learning at the college level, that term typically refers to internships or lab classes, and you are fairly certain it doesn't apply to mathematics. So you check the sign on the door, and sure enough, this is the correct math class, taught by any one of the contributors to this book, including the two of us.

The value of active learning, which includes tactile learning, is well documented in the literature. The benefits of these active learning approaches include increased accessibility, deeper understanding, and a way to level the playing field for students with diverse backgrounds and levels of mathematical preparation [3, 5, 7, 8]. In addition to the strong evidence in the literature, both of us have had the joy of observing our students dissect, discuss, distill, and discover mathematics through tactile learning. We encourage you, the curious instructor, to look through the literature to get a clearer understanding of how activities can enhance learning in your classes.

The two of us have been using a variety of tactile activities in our classes for years, and we are constantly seeking and developing new ideas. We co-organized an MAA Special Session, "Touch it, feel it, learn it: Tactile learning activities in the undergraduate mathematics classroom," for the 2012 JMM in Boston. With 35 presentations drawing an audience of 100-150 people, we decided to offer a second MAA Special Session at the 2013 JMM in San Diego, where we had an additional 26 well-attended presentations. Given the diversity of the presentations at these sessions, it is clear that there a a large number of mathematics faculty members using a wide variety of tactile activities in their classes, and an even larger number of people showing interest in the subject.

In response to the high level of interest from the mathematical community, we guest-edited a special issue of *PRIMUS* [2] dedicated to hands-on activities in which the authors, many of whom have also contributed to this book, offered specific pedagogical support for the activities presented therein. We have also assembled this book, offering a diverse collection of activities targeting mathematical topics ranging in level from precalculus to knot theory. We developed this book for you, the curious instructor—both the seasoned experimentalists and those trying hands-on activities for the first time, and we hope that you find it beneficial as you invite your students to explore mathematics tactilely in your classrooms.

We are sure that all of the people who spoke in our sessions, attended the talks, authored articles in our *PRIMUS* issue, or contributed to this book have their own tales about what initially sparked their interest in teaching this way. We include our stories below.

Jessica's story: My calculus class at West Point was just getting ready to transition from single variable integration to integrating functions of two variables, and I knew that my students would struggle with the concept of "the volume beneath a surface". I spent my entire afternoon writing, scrapping, re-writing, and re-scrapping lesson plans, Mathematica notebooks, and chalkboard sketches to make the concept accessible, but nothing seemed concrete enough. The night before the lesson, I went to bed very worried. As I sleeplessly tossed and turned, I caught sight of my arm moving under the sheet, creating a beautiful surface. I sat straight up, realizing that my students could stand in formation, and by tossing a sheet over them, my class could create their own surface, using their own heights to approximate the integral. I tried this activity in class the next day, and based on the positive feedback from my students and the prodding of a colleague, I wrote an article on this activity that appeared in *PRIMUS* [6].

The year after developing the sheet activity for my calculus class, I was asked to develop a new course called Mathematics for Space Applications; there was no similar course in the country at the time. Drawing students from systems engineering, mathematics, physics, and a handful of other majors, I had difficulty designing lectures that addressed the broad backgrounds of my students. Shortly into the course, I tossed out the majority of my lectures and went to a full peer-teaching and hands-on learning approach. We represented orbits using hula hoops and globes; we made model solar systems with sidewalk chalk and string; we spun ourselves silly in office chairs as we traced out highly elliptical orbits with our feet; and most importantly, we had fun while we learned. I wrote an article on the format of this course, including the hands-on activities, for *Mathematica Militaris* [4]. Through teaching this course, I developed a new appreciation for hands-on activities as a way of leveling the playing field for a diverse group of students, and I continue to leverage this benefit in my teaching at the Virginia Military Institute.

Julie's story: After covering the epsilon-delta definition of continuity in real analysis, I noticed many pairs of glazed eyes staring back at me. The definition had been too abstract for the students, so I tried to explain the definition again. I drew a diagram, but it was too static. I used technology to zoom in on a graph, but students only saw the lines instead of connecting them to the original function being studied. We were all frustrated. A few days later while walking down the craft aisle in a store, I stumbled on a feather boa that had haphazardly fallen onto the floor. The way it was situated, the boa resembled a large graph of a function, and I could imagine my students walking on it while physically exploring that analysis definition. Later in class, my students were surprised when I placed a feather boa on the classroom floor. Students volunteered to use yarn to represent the epsilon and delta regions around a point. They noticed how delta was affected by epsilon. The definition came alive. Throughout the rest of the course, we used the boa or a collection of boas whenever we covered new definitions about functions. Details for that activity can be found in *PRIMUS* [1].

After successfully using feather boas in analysis, I realized that the boas could also be used in any class that studies graphs of simple functions. I gathered a collection of feather boas and developed small group activities to aid students in understanding piecewise functions as well as activities for making connections between derivatives and the shape of graphs. The added beauty of using the boas in calculus and precalculus is that students typically are intrigued by the novelty and want to touch them; this increases student participation. Also, the graphs are large enough to see from the other side of the room, making it easy to tell which groups understand the material and which groups need a little more help. Since the boas worked well for small group activities, I brought them into my senior complex variables class where students used the boas to represent the images of a basic smiley face under a variety of standard complex functions. I never realized that feather boas were such a great teaching tool in mathematics. Even more, I never realized how an idea that worked in one class could be modified to create useful activities in a wide range of classes.

We have both realized the value of hands-on teaching and learning and have adopted it in a wide array of applications. Whether presenting a challenging concept or leveling the playing field for a diverse student population, we try to develop a meaningful hands-on activity that forces our students to engage with the material, ultimately developing their own concrete understanding. Of course, the joy of seeing a room full of engaged and excited students comes at a price, as these activities require planning and take time away from lecturing. Yet, activities do not have to be done at the expense of content. Lecture time is simply replaced with the activity, and the class then has a shared experience that can be referenced for many lessons down the road. Therefore, we believe that the benefits of using activities in class are worth the investment.

Each of the activities in this book has been used and vetted by its author(s), and the write-up includes suggestions and pitfalls to help reduce that initial investment of time. A more detailed explanation of the features of this book can be found in "How to Use This Book". We invite you to try some of the activities in this book, and we hope that you and your students both benefit from the deeper learning and the simple joy that hands-on activities can bring.

Sincerely,
Julie & Jessica

References

1. J. Barnes, Feather boas in real analysis, *PRIMUS*, 21 no. 2 (2011) 130–141.

2. J. Barnes and J. Libertini, Special Issue on Tactile Learning Activities, *PRIMUS*, 23 no. 7 (2013).

3. S. Freeman, S. L. Eddy, M. McDonough, M. K. Smith, N. Okorafor, H. Jordt, and M. P. Wenderoth, Active learning increases student performance in science, engineering, and mathematics, *Proc. Nat. Acad. Sci. (PNAS)* 111 no. 23 (2014) 8410–8415.

4. J. Libertini, A new course for a new mission: mathematics for space applications, *Mathematica Militaris* 20 no. 3 (2012) 13–19.

5. J. Michael, Where's the evidence that active learning works? *Advances in Physiology Education* 30 (2006) 159–167.

6. J. Mikhaylov, Be the volume: A classroom activity to visualize volume estimation, *PRIMUS*, 21 no. 2 (2011) 175–182.

7. M. Prince, Does active learning work? A review of the research, *J. Engineering Education* 93 no. 3 (2004) 223–231.

8. C. E. Weiman, Large-scale comparison of science teaching methods sends clear message, *Proc. Nat. Acad. of Sci. (PNAS)* 111 no. 23 (2014) 8319–8320.

How to Use This Book

Consider using this book in much the same way you would use a cookbook. Each hands-on learning activity is presented in recipe format with generally two pages describing the reasons, logistics, and helpful hints for running the activity, and another page that can be used as a handout in class.

> These handouts are available free of charge at:
> `www.ams.org/bookpages/clrm-54`.

The topics presented cover a variety of mathematical concepts found in courses at all levels. The activities are grouped by level. Activities designed for use in courses before calculus are presented as appetizers, as these courses provide an early introduction to the field of mathematics. Calculus activities are referred to as main courses, as the calculus sequence is a substantial portion of any mathematics curriculum. The book finishes with activities that are relevant to upper-level courses; these activities are called desserts given the joy that is often presented and discovered in these courses. Although the focus of this book is on courses taken by mathematics majors, several of the activities in this book can also be used in lower level exploratory topics courses, such as Math for Liberal Arts; an overview of such activities is presented on pages xvi-xx.

In terms of the time commitment needed to run these activities, they range from tiny morsels, short activities that require only a few minutes for implementation, to more hefty portions, long activities that would need to be savored for an entire class period.

As with any cookbook, it is possible to flip to an activity and follow the instructions exactly as presented. However, just as you might modify a recipe based on your taste or the availability of ingredients, we encourage you to modify the details of any activity to meet the specific needs of your students and constraints of your course. We also invite you to browse through the book simply for inspiration as you create and implement your own activity ideas.

We have included several indices to help you navigate the book.

Concept Index: This index makes it possible to search for activities related to a particular mathematical concept.

Author Index: This index can be used if you liked one activity and would like to see more by that author.

Main Ingredient Index: This index can be used to find an activity that uses a particular prop, like cookies, a sheet, or feather boas. It could also be used to see if this book includes an activity that describes mathematics using a random item you found on your shelf.

Course Index: This index identifies all the activities that could be used in a specific mathematics course. Activities may be listed under multiple courses.

We hope that this book will encourage and inspire you to explore the possibilities of using more hands-on activities in your classes. Bon appétit!

Modifications for Liberal Arts and Topics Courses

General education courses that explore a wide variety of mathematical topics, with titles like Mathematics for Liberal Arts, are becoming increasingly common. Many of the activities in this book can be used in such a course either as written or with modifications. Tables 1-6 give a brief overview of some topics and the activities that address them. Many of these activities do not require modification and are denoted by "-" in the corresponding comment column. Suggestions or modifications for all other activities are discussed in the numbered list below each table, where the numbers correlate to the notes column in the table.

Functions

Many lower level courses introduce students to functions. Four of these activities may require modifications that are given in the notes below. However, most of the activities listed in Table 1 on this topic can be used without modification and are designated with a "-". Regardless of whether modifications are discussed for a particular activity, we encourage you to think about your course and tailor that activity accordingly.

Table 1: Function activities that would work in a liberal arts course.

Activity Number	Activity Name	See Note Number
1.2	Function Ball Toss	-
5.13	Properties of Functions on Finite Sets Using Candy	1a
1.4	Function Composition Using Crackers and Cheese	-
1.5	Walking Function Transformations	-
1.1	Using Parenthesis with the Game of Telephone	-
1.6	Graphing Piecewise Functions with Feather Boas	-
4.2	Building Functions of Two Variables with Cookies	1b
4.3	Exploring Contours in the Physical World	1b
4.4	Matching Photographs with Contour Lines	1b

Notes:

1a. **Properties of Functions on Finite Sets Using Candy**: This activity, which can help introduce students to the definition of a function, was originally designed for an upper level course. As written, the activity has students engage with some of the more formal terminology, such as *injection*, *surjection*, *bijection*, *one-to-one*, and *onto*, and the handout uses this terminology as well as formal notation. Therefore, the handout may need to be altered to fit your course.

1b. **Building Functions of Two Variables with Cookies**, **Exploring Contours in the Physical World**, and **Matching Photographs with Contour Lines**: Each of these activities allows students to explore multivariable functions and methods of visualizing these functions at an introductory level. As these activities were originally designed for use in a multivariable calculus course, you may want to avoid using calculus terms in your explanations. On the handouts, you may opt to remove the formal domain notation used in **Cookies**; the other two handouts may be used as written.

Modeling

This book contains eight activities that relate to mathematical modeling, or the contextualization of mathematics. Although many of these activities were originally developed for the calculus level or higher, each can be made accessible to a broader audience through the suggestions given below.

Table 2: Modeling activities that would work in a liberal arts course.

Activity Number	Activity Name	See Note Number
1.3	Mathematical Modeling Using Long-exposure Photography	2a
2.1	Adding Movement to Velocity Explorations with Ziplines	2b
2.6	Spread the Word: Modeling Logistic Growth	2c
2.7	Maximizing the Area of a Fenced in Region Using...	2d
2.8	The Optimal Origami Box	2e
5.3	Population Modeling Using M&M's	2f
5.4	Modeling of Fishing and Restocking with Pennies	2f

Notes:

2a. **Mathematical Modeling Using Long-exposure Photography**: This activity is already targeted for a lower level course. If you wish to increase the accessibility further, remove the portion on computing error (Question 4 on the handout).

2b. **Adding Movement to Velocity Explorations with Ziplines**: This activity is designed to be used early in a calculus course as an exploratory introduction to the ideas of a limit and ultimately a derivative. As it is an introductory activity, the handout already avoids calculus terminology such as *limit* and *derivative*. Therefore, the handout for this activity can be used as is; we just caution you to avoid using calculus terminology in the discussion.

2c. **Spread the Word: Modeling Logistic Growth**: In this activity, students collect data on the spread of a disease amongst their classmates and do a fit to the data. Question 6 asks students to perform a logistic regression, which can be done using technology, avoiding the need for knowledge of this skill. Question 7 refers to an inflection point; if you choose to maintain the use of this terminology, you may need to define and discuss this with your students before they attempt that question.

2d. **Maximizing the Area of a Fenced in Region Using Bendable Sticks for Constraints**: Although this activity in optimizing a fenced area was originally written for a calculus course, most of the handout does not require any knowledge of calculus. Question 6 asks students to use calculus to find the optimal area, but this can be replaced with a graphical method without further modification.

2e. **The Optimal Origami Box**: This activity was originally designed for use in the optimization section of a calculus course, but a graphical optimization technique could be used instead. The handout is already written without reference to calculus ideas, as students are asked to develop their own approach to optimization. Therefore, the handout can be used as written, and you can simply direct student discussion towards a graphical approach.

2f. **Population Modeling Using M&M's** & **Modeling of Fishing and Restocking with Pennies**: Both of these activities are designed to motivate the study of differential equations through experiments in population dynamics. Although the instructor guidance for both drive these activities towards developing differential equations, the process of understanding the experiments, developing hypotheses, and testing these hypotheses are all still accessible without using a differential equation. Restricting the development to difference equations keeps the mathematics accessible, as these can be easily analyzed by hand or using a spreadsheet. The handout for **M&M's** is already written for a general audience, as it does not discuss differential equations. On the handout for **Fish**, you may wish to tailor Questions 3c and 4, as these questions currently deal with the formalization of the mathematical model, the solution process, and the assessment of the model.

Trigonometry

If your topics course includes a trigonometry component, then you may find two of the activities in this book helpful, one of which may require a minor modification.

Table 3: Trigonometry activities that would work in a liberal arts course.

Activity Number	Activity Name	See Note Number
1.7	Trigonometry Parameter Comparisons	-
4.12	Vector Analysis of Pop-Up Page	3a

Notes:

3a. **Vector Analysis of a Pop-Up Page**: This activity, which has been used in an introductory Math and Art course, provides an opportunity to apply trigonometry in three dimensional coordinate systems, which need to be taught prior to this activity. On the handout, Question 7 asks students to set up equations relating the positions of points of the pop-up. As this can be challenging, it is advisable that you demonstrate the first part of Question 7 at the board before groups attempt the rest of the question. It may even be necessary to walk the class through the entire problem, depending on the algebra skills of your students. All other portions of the activity should be accessible as long as your students understand algebraic manipulation and you've introduced basic trigonometry and three dimensional coordinate systems.

Logical Reasoning

Understanding logical arguments is an important and transferable skill, and as such, logical reasoning is a common topic in a general education topics course. All of the activities listed in the table can be used without modification, but we encourage you to look over the terminology and notation used on the handouts to make sure that you introduce these to your students as needed.

Table 4: Logic activities that would work in a liberal arts course.

Activity Number	Activity Name	See Note Number
5.8	Traffic Jam: A Lifesize Logic Puzzle	-
5.9	Living DeMorgan's Laws	-
5.10	Using Circuits to Teach Truth Tables	-
5.11	Determining the Validity of an Argument Using...	-
6.8	Walking the Seven Bridges of Konigsberg	-

Approximating Rates of Change, Areas, Volumes, and Series

Although most students in a general education level topics course will not take calculus, the ideas of calculus have broad applications, and it may be beneficial for all students, regardless of major, to have exposure to these ideas. Many of the exploratory calculus activities in this book can be easily modified to help all students understand the underpinning concepts of calculus without actually doing any calculus.

Table 5: Approximating rates of change, areas, volumes, and series activities that would work in a liberal arts course.

Activity Number	Activity Name	See Note Number
2.1	Adding Movement to Velocity Explorations with Ziplines	5a
3.1	Chewing Gum Riemann Sums	5b
3.2	Riemann Sums Using the Paper Shredder or Cut the Bunny	5b
3.3	Estimating Calories in a Cookie with Riemann Sums	5b
3.7	Fun with Infinite Series	5c
4.8	Volume Estimation Using a Sheet Surface	5d
4.9	Visualizing and Estimating the Mass of a Solid Using...	5d

Notes:

5a. **Adding Movement to Velocity Explorations with Ziplines**: This activity encourages students to think about how one might determine instantaneous rates of change based on experiments calculating average rates of change over a decreasing distance. This activity motivates the idea of a derivative without ever actually using that terminology. The handout can be used as written, and you should just be mindful of the terminology you use during the activity.5b. **Chewing Gum Riemann Sums, Riemann Sums Using the Paper Shred-**

der or Cut the Bunny, and **Estimating Calories in a Cookie with Riemann Sums**: Each of these activities has students explore area estimations using Riemann sums. None of the handouts use the terms relating to integration, so they may be used as written, unless you also wish to avoid the terminology of Riemann sums, in which case, the **Bunny** activity may be your preferred choice as it is very open-ended and does not use any formal language to guide the inquiry process.

5c. **Fun with Infinite Series**: This title actually covers three activities relating to infinite series, two of which are easily adapted for a lower level course. The **Fun with Paper** activity could be used without modification. The **Fun with Fractals** activity could be modified to remove the formal notation and calculation of series. Both of these activities can be used to reinforce algebraic notation and geometric sequences as students consider the sums and make hypotheses about whether the results are finite or infinite. The handouts are written in such a way that they can be used without modification. The **Fun with Gravity** activity could be similarly modified, but it may be more challenging for students and requires some understanding of physics.

5d. **Volume Estimation Using a Sheet Surface** and **Visualizing and Estimating the Mass of a Solid Using Multi-colored Blocks**: These two activities have students explore simple approximations for the volume and mass, respectively, of a three dimensional object. The **Sheet** activity handout can be used as written, as it does not use any formal terminology or notation. You may wish to modify the handout for the **Blocks** activity, as it uses formal notation.

Exposure to Higher Mathematics

Several topics courses are designed to introduce students to some interesting ideas found in higher level mathematics that they would otherwise not encounter in their college careers as non-mathematics majors. The activities listed in this section of the table offer opportunities to support this type of course. Many of these activities do not require modification; those that require modification are discussed below. For all activities, we encourage you to be mindful of the terminology and notation you introduce and use in your course.

Table 6: Higher level mathematics activities that would work in a liberal arts course.

Activity Number	Activity Name	See Note Number
6.8	Walking the Seven Bridges of Königsberg	-
6.9	Designing Round - Robin Tournaments Using Yarn	-
6.15	Exploring Knots	-
6.1	Using Candy to Represent Equivalence Relations	-
5.7	Picturing Prime Factorization	-
6.7	Discovering Catalan Numbers Using M&M's	-
6.3	Symmetry and Group Theory With Plastic Triangles	6a
6.5	Acting Permutations	6b
5.14	SET in Combinatorics and Discrete Math	6c

Notes:

6a. **Symmetry and Group Theory With Plastic Triangles**: This activity, which focuses on identifying groups, can be used in a topics course to demonstrate an abstract side of mathematics that highlights the field as one of the liberal arts. The questions on the handout get at some deeper concepts, including identities and inverses, which could be excluded; however, much of this activity, including the building of the group table has been completed by children in the fourth grade.

6b. **Acting Permutations**: This activity has been used in both mathematics education and mathematics for liberal arts courses by modifying the activity to limit the scope of the explorations to permutations, inverses, commutativity, identity, and order. To add meaning to the activity, you can emphasize examples of permutations in the real world and the thought process behind generating and testing conjectures.

6c. **SET in Combinatorics and Discrete Math**: This activity can be used in a wide variety of settings, as long as you introduce your students to combinations and the multiplication principle. Students in a lower level course may have difficulty with Question 2 on the handout, so you may consider removing it or providing students with more guidance than you would in an upper level mathematics course.

Part I

Appetizers (Before Calculus)

Chapter 1. Precalculus

Chapter 1

Precalculus

1.1 Using Parentheses with the Game of Telephone
1.2 Function Ball Toss
1.3 Mathematical Modeling Using Long-Exposure Photography
1.4 Function Composition Using Crackers and Cheese
1.5 Walking Function Transformations
1.6 Graphing Piecewise Functions with Feather Boas
1.7 Trigonometry Parameter Comparisons

1.1 Using Parentheses with the Game of Telephone

Julie Barnes, Western Carolina University
Jessica Libertini, Virginia Military Institute

Concepts Taught: parentheses, order of operations, mathematical notation

Activity Overview

Most college students have been taught the order of operations and proper usage of notation, especially parentheses, well before coming to college. However, many students seem to have forgotten those rules and see no need for parentheses, writing out steps, or carrying through limits. This activity shows students the importance of correctly using proper mathematical notation. It is a lot like the childhood game of Telephone where the first child is given a message, whispers it to the next child, and after passing the message to 20-30 children in this fashion, the message is often distorted. Here, instead of whispering, each student works one step of the problem in isolation and passes the result to the next person, who in turn does the next step, and so on. If everyone uses proper notation and simplifies the mathematical expressions correctly, the "mathematical message" is transmitted without error. If parentheses are used incorrectly or any other notation errors are made, the answer that appears on the other end will most likely be incorrect. Students seem more willing to believe that the details of mathematical notation, such as parentheses, are necessary when their fellow students misinterpret their work rather than an instructor who may be perceived as being too picky.

Supplies Needed (per group)	Class Time Required	Group Size
Prepared problem* Prepared stack of sticky notes** roughly 1 sticky note per person	10 minutes	5-10 students depending on the problem

*This activity works best with each group doing the same problem at the same time. You could choose one from the four sample problems provided on page 7 with possible solutions on pages 8 and 9, but you are encouraged to choose a problem of your own that addresses any specific issues your students have encountered. Copy the problem you want your students to solve on several sticky notes, one sticky note per group.
**For each group, you will also need a stack of 5-10 sticky notes depending on the number of steps needed to simplify the problem chosen. Stick one of the prepared problems on the top of each stack. Add a few extra sticky notes to the bottom of the stack just in case your students do more steps than you had anticipated. It is helpful if each stack of sticky notes is a different color for future reference.

Running the Activity

Explain the game of Telephone and tell students that they will be doing one step toward solving a problem and presenting a final answer in simplified form. Hand one student per group a stack of sticky notes complete with a problem on the top sticky note. Tell that student to remove the top sticky note, stick it to his or her desk, and leave it there for the remainder of the activity. Then the student should write down the first step needed to evaluate or simplify the problem on the top sticky note remaining. After recording the first step, the student should pass the stack of sticky notes to another student in his or her group. The next student removes the top sticky note which has the previous students' work on it, sticks it to his or her desk, and leaves it there for the remainder of the activity. Then, he or she does the "next step" in simplifying the expression and writes it on the top sticky note of the remaining stack. Students continue this process until they feel the expression is simplified as far as it can be simplified.

Once everyone has finished, have students share their answers with the class. This activity is most helpful if some mistakes are made along the way and not all answers are the same. Assuming that not all answers are the same, ask students what happened and give them a chance to determine what mistakes were made. For the samples provided, common mistakes include leaving off the limit sign so that the answers still contain a variable, losing parentheses, and incorrectly substituting items into functions. Possible correct solutions to the sample problems are provided on pages 8 and 9; these could be projected via a document presenter in class to aid in the class discussion once students have determined what mistakes they think were made. In the rare event that all problems are solved perfectly, congratulate them for their accomplishment, and lead a discussion on why the mathematical ideas were transmitted without errors. In all cases, be sure that students walk away with the notion that details in mathematical notation are important.

Suggestions and Pitfalls

Make sure that students are not helping the student either before or after his or her step. That would eliminate the need for proper mathematical notation.

This activity also works well if pairs of students work on each step instead of individual students. That is, a pair of students would work on the first step, and pass their work to a different pair of students, and so on. Working in pairs is particularly helpful for students who are afraid of mathematics.

In lieu of a handout, you may want to ask your students to write a short paragraph about the importance of notation in mathematics.

Sample Problems

A. Simplify $-3(2-4)^2 + 24 \div 3 \times 2$.

B. Expand and simplify $-3x(x+1)(x-2)^2$.

C. Simplify $\dfrac{\frac{1}{x} - \frac{x+1}{x^2}}{x^3}$.

D. Simplify $\dfrac{\frac{1}{x-b} - \frac{1}{x}}{b}$.

E. Evaluate and simplify $\lim_{h \to 0} \dfrac{f(x+h) - f(x)}{h}$ where $f(x) = x^2 - 5$.

F. Let $f(x) = x^2 - 3x + x^3, g(x) = 1 - x$, and $h(x) = e^{-x}$.
Evaluate and simplify $f(-1)(g(2+e) + h(0))$.

G. Evaluate and simplify $\lim_{x \to 5} \dfrac{\frac{2}{x^2} - \frac{2}{25}}{x - 5}$.

H. (*for calculus*) Evaluate and simplify $\dfrac{d}{dx}\left((x^2 + x)^3\right)$.

Possible Correct Solutions to Sample Problems

A.

$$\begin{aligned}-3(2-4)^2+24\div 3\times 2 &= -3(-2)^2+24\div 3\times 2\\ &= -3(4)+24\div 3\times 2\\ &= -3(4)+8\times 2\\ &= -3(4)+16\\ &= -12+16=4.\end{aligned}$$

B.

$$\begin{aligned}&-3x(x+1)(x-2)^2\\ &\quad=(-3x^2-3x)(x-2)^2\\ &\quad=(-3x^2-3x)(x^2-4x+4)\\ &\quad=-3x^4+12x^3-12x^2-3x^3+12x^2-12x\\ &\quad=-3x^4+9x^3-12x.\end{aligned}$$

C.

$$\begin{aligned}\frac{\frac{1}{x}-\frac{x+1}{x^2}}{x^3} &= \frac{\frac{x}{x^2}-\frac{x+1}{x^2}}{x^3}\\ &= \frac{\frac{x-(x+1)}{x^2}}{x^3}=\frac{\frac{x-x-1}{x^2}}{x^3}\\ &= \frac{\frac{-1}{x^2}}{x^3}=\frac{-1}{x^2}\cdot\frac{1}{x^3}\\ &= \frac{-1}{x^5}.\end{aligned}$$

D.

$$\begin{aligned}\frac{\frac{1}{x-b}-\frac{1}{x}}{b} &= \frac{\frac{x}{x(x-b)}-\frac{x-b}{x(x-b)}}{b}\\ &= \frac{\frac{x-(x-b)}{x(x-b)}}{b}=\frac{\frac{b}{x(x-b)}}{b}\\ &= \frac{b}{x(x-b)}\cdot\frac{1}{b}=\frac{1}{x(x-b)}\\ &= \frac{1}{x^2-bx}.\end{aligned}$$

E. $$\lim_{h\to 0}\frac{f(x+h)-f(x)}{h} \qquad \text{where } f(x)=x^2-5$$

$$\begin{aligned}
&=\lim_{h\to 0}\frac{(x+h)^2-5-(x^2-5)}{h}\\
&=\lim_{h\to 0}\frac{x^2+2xh+h^2-5-(x^2-5)}{h}\\
&=\lim_{h\to 0}\frac{x^2+2xh+h^2-5-x^2+5}{h}\\
&=\lim_{h\to 0}\frac{2xh+h^2}{h}=\lim_{h\to 0}\frac{h(2x+h)}{h}\\
&=\lim_{h\to 0}2x+h=2x.
\end{aligned}$$

F. $$\begin{aligned}
\lim_{x\to 5}\frac{\frac{2}{x^2}-\frac{2}{25}}{x-5}&=\lim_{x\to 5}\frac{\frac{50-2x^2}{25x^2}}{x-5}\\
&=\lim_{x\to 5}\frac{50-2x^2}{25x^2}\cdot\frac{1}{x-5}=\lim_{x\to 5}\frac{2(25-x^2)}{25x^2(x-5)}\\
&=\lim_{x\to 5}\frac{2(5-x)(5+x)}{25x^2(x-5)}=\lim_{x\to 5}\frac{-2(x-5)(5+x)}{25x^2(x-5)}\\
&=\lim_{x\to 5}\frac{-2(5+x)}{25x^2}=\lim_{x\to 5}\frac{-20}{625}=\frac{-4}{125}.
\end{aligned}$$

G. $$f(x)=x^2-3x+x^3,\quad g(x)=1-x,\quad h(x)=e^{-x}$$

$$\begin{aligned}
f(-1)(g(2+c)+h(0))&=((-1)^2-3(-1)+(-1)^3)(1-(2+c)+e^{-0})\\
&=(1+3-1)(1-(2+c)+e^{-0})\\
&=3(1-(2+c)+1)\\
&=3(2-(2+c))\\
&=3(2-2-c)\\
&=3(-c)\\
&=-3c.
\end{aligned}$$

H. $$\begin{aligned}
\frac{d}{dx}\left((x^2+x)^3\right)&=3(x^2+x)^2(2x+1)\\
&=3(x^4+2x^3+x^2)(2x+1)\\
&=(3x^4+6x^3+3x^2)(2x+1)\\
&=6x^5+12x^4+6x^3+3x^4+6x^3+3x^2\\
&=6x^5+15x^4+12x^3+3x^2.
\end{aligned}$$

1.2 Function Ball Toss

Julie Barnes, Western Carolina University

Concepts Taught: functions, inverse functions, function tables

Activity Overview

Students in introductory college mathematics classes typically have seen graphs and equations of functions, but are not always comfortable with what it actually means to be a function. In this activity, students pass balls to each other as described by a table of data. In some cases, the rules for passing the balls represent a function and sometimes they do not. Students are able to physically see the difference between the two based on whether the rules are clear or not, giving them a concrete picture of what it means to be a function.

Supplies Needed	Class Time Required	Group Size
4 identical balls	10 minutes	Class demonstration

Running the Activity

In order to get students thinking about functions, have them imagine a candy machine. You put money in the machine and push a button. If the machine is working properly, exactly one item comes out; the machine is therefore a function. If two buttons yield the same type of candy, the machine is still a function because each button produces one result. However, if one button yields two types of candy, the machine is not working properly, and consequently, it is not a function.

Provide students with a copy of the handout to guide the discussion, and ask for four volunteers to stand in front of the room. Call the four people A, B, C, and D, and give each one a ball. Tell your class that the volunteers will be passing their balls according to specific rules provided from different tables. The job of each volunteer is to find his/her letter in the left column of a given table as well as the corresponding letter in the right column. On the count of three, each volunteer passes his/her ball to the person whose letter is in the right column. In Table 1, volunteers pass balls according to $f(x)$: A passes the ball to B, B passes the ball to C, C passes the ball to D, and D passes the ball to himself/herself. Figure 1.1 shows students acting out the description in Table 1. After the quick demonstration, give students a chance to complete Problem 1 from the handout and share their responses with others nearby. Then briefly discuss their answers as a class.

Once students realize that $f(x)$ from Table 1 is a function, have the volunteers reset such that each person has one ball, and then repeat the process with Table 2. Now, $g(x)$ is the rule that says A passes the ball to B, B passes the ball to C, C passes the ball to A, D passes the ball to A, and D passes the ball to D. When students are asked to pass their balls, D should notice quickly that he/she cannot do two things at once with the same ball. Have students complete Problem 2 and then share their responses with the class. They should notice that $g(x)$ is not a function because one element of the domain has unclear directions. Figure 1.2 shows students acting out the description in Table 2.

Figure 1.1: Students passing the ball according to Table 1 from the handout.

Figure 1.2: Students passing the ball according to Table 2 from the handout. Notice that Table 2 does not represent a function because Student D is not able to do what the table asks him to do.

Again have the volunteers reset so that each person has one ball. Have them act out Table 3 where the function $h(x)$ is the rule that says A passes the ball to B, B passes the ball to C, and C passes the ball to A. This leads to an interesting observation. D is standing there holding a ball and has no instructions. Have students complete Problem 3 and share answers with the class. Help them realize that this is not a function on all of the volunteers, $\{$A, B, C, D$\}$, but it is a function on the smaller domain of just $\{$A, B, C$\}$.

Finally, repeat the function ball toss described in Table 1. This time, have students respond to Problem 4 from the handout. Tie the notion of invertible functions to whether each person in the range knows who tossed the ball(s) to him/her. If a ball could have come from more than one source, then the function is not invertible.

Depending on the needs of your students, you could include more tables and/or have a longer discussion about what makes a process a function or not.

Suggestions and Pitfalls

Often, the hardest part is getting volunteers if your students are afraid of mathematics. It may be useful to emphasize that the volunteers only need to have basic ball tossing skills.

Note that we are using function notation in the tables even though some of the tables do not represent functions. If this is a concern, you may want to point this out to your students. Also, this activity could be modified to address other topics, such as determining if a function is one-to-one or onto.

Function Ball Toss – Class Handout

Table 1

x	$f(x)$
A	B
B	C
C	D
D	D

Table 2

x	$g(x)$
A	B
B	C
C	A
D	A
D	D

Table 3

x	$h(x)$
A	B
B	C
C	A

1. After the four volunteers demonstrate the action described by Table 1, complete the following problems.

   (a) A is not mentioned in the second column. What is unique about A's interaction with the ball toss? Does this keep $f(x)$ from being a function?

   (b) D is listed twice in the $f(x)$ column. What was different about D's interaction with the ball toss? Does this keep $f(x)$ from being a function?

   (c) Did all the volunteers know where they were supposed to toss the ball?

   (d) Does $f(x)$ represent a function? Why or why not?

2. After the four volunteers demonstrate the action described by Table 2, complete the following problems.

   (a) A is listed twice in the $g(x)$ column. What was unique about A's interaction with the ball toss?

   (b) D is listed twice in the x column. What was unique about D's interaction with the ball toss?

   (c) Did all the volunteers know where they were supposed to toss the ball?

   (d) Does $g(x)$ represent a function? Why or why not?

3. After the four volunteers demonstrate the action described by Table 3, complete the following problems.

   (a) Did all the volunteers know where they were supposed to toss the ball?

   (b) Does $h(x)$ represent a function on the domain $\{$A, B, C, D$\}$? Why or why not?

   (c) Is $h(x)$ a function on a different domain? Why or why not?

4. After the four volunteers repeat the procedure described in Table 1, complete the following problems.

   (a) Does B know who tossed the ball to him/her?

   (b) Are there any volunteers who received balls from multiple people?

   (c) If $f(x)$ is invertible, explain why; if not, present a modified version of $f(x)$ such that this new version is invertible.

1.3 Mathematical Modeling Using Long-Exposure Photography

Johann Thiel, New York City College of Technology

Concepts Taught: modeling, rate of change, linear functions, curve fitting

Activity Overview

An important and useful skill in multiple fields is the ability to construct appropriate mathematical models based on collected data. The goal of this activity is to create unique in-class data sets that students can then use to practice mathematical modeling. In particular, as this is an introductory lesson into the topic, we will focus on deciding whether or not a linear model is appropriate for a given data set. The instructor and students will use a digital camera and a flashing rubber ball to create data sets of the ball's position as a function of time. By setting the camera to a very slow shutter speed, the long-exposure images created will capture the motion of the flashing rubber ball in discrete steps.

Supplies Needed	Class Time Required	Group Size
Digital camera* Tripod Computer & projector Meter stick Flashing rubber ball**	15-20 minutes	2-3 students

* The camera must have an adjustable shutter speed. Some smartphones are equipped with this feature.
** Flashing rubber ball toys can be purchased in pet stores, Walmart, Target, etc.
Note: We recommend that you test the camera and upload one or two shots onto the computer before class starts. This will help reduce the chances of technical issues arising in the middle of the activity. Also, before starting class, the flashing rubber ball should be photographed while in motion with the camera's shutter speed set to between 1-2 seconds. This is to help approximate the time interval between flashes, which can be computed as the shutter speed (in seconds) divided by the number of light points that appear in the photograph. For example, Figure 1.3b shows 23 flashes over an 8 second exposure, resulting in a period of approximately $\frac{1}{3}$s. The time period between flashes will be needed by the students in order to complete the handout.

Running the Activity

To begin the activity, set up the camera on the tripod facing a designated student holding the flashing rubber ball. Place the meter stick in a location that is as far away from the camera as the student. See Figure 1.3a for an example of the setup. Set the camera's shutter speed to at least four seconds, and dim the lights in the room.

For the first data set, photograph the movement of the flashing rubber ball as it rolls across the floor in front of the camera. For the second data set, photograph the movement of the flashing rubber ball as it is dropped from a height of at least five feet. The goal is to drop the ball with a slight horizontal velocity. This will make it easier to measure the ball's position as it falls for the first time. See Figures 1.3b and 1.3c for examples of resulting images.

(a) Example setup with meter stick.

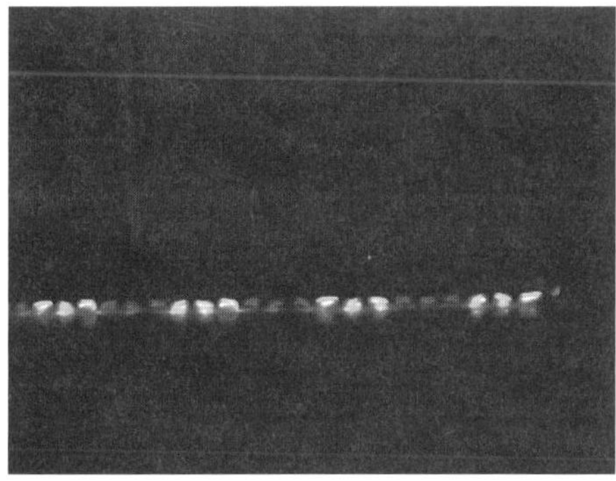

(b) Flashing rubber ball rolling on the floor.

(c) Dropping the flashing rubber ball.

Figure 1.3: Long-exposure images of the flashing rubber ball.

Project the resulting images onto a screen that everyone can see. Have the class work together to measure and record the position of the ball relative to a fixed point for each flash. The class should now be able to start on the exercises in the handout. The handout guides students through the process of tabulating the data obtained from the images, graphing it, and answering questions on the use of appropriate mathematical models to describe the movement of the ball.

Once students are nearly finished with the handout, begin discussing some of the differences observed in the images. Students should be able to explain if they believe that a linear model suits either the rolled ball or the dropped ball. In the cases where a linear model appears to be a good fit, follow-up questions on how to construct the model should be pursued. In practice, this can lead to the introduction of derivatives and the acceleration due to gravity.

Suggestions and Pitfalls

For this lesson, when we discuss the position of the ball, we are referring to only one direction at a time. For example, in Figure 1.3b, position refers to the horizontal distance (x-position) of the ball over time, while in Figure 1.3c, position refers to the vertical height (y-position) of the ball over time. Depending on the case, when measuring the ball's position relative to a fixed point, we mean strictly either the horizontal or vertical position, but not both.

The position of the ball relative to a fixed point on the projected image can be measured using the meter stick by simply holding it up to the projected image. As it is unlikely that the projected image is to scale, the photographed meter stick from Figure 1.3a will help with converting the measured units to the appropriate scale.

If there is sufficient time, individual groups can make their own measurements. This can lead to interesting discussions about how to properly measure the position of the ball at any point in time. If such time is not available, you can still offer students the opportunity to take part in the process by having each come to the board to perform one position measurement that the entire class will use.

Mathematical Modeling Using Long-Exposure Photography – Class Handout

1. Fill out the tables below using the measurements collected in class. The columns are t for time, Δx and Δy for the changes in position over each time interval, and x and y for the positions relative to the starting point at time t.

t	Δx	x

Table 1.1: Rolled ball.

t	Δy	y

Table 1.2: Dropped ball.

2. Plot the position versus time (x vs. t) and (y vs. t) for each table on the space below. Be sure to label the axes appropriately.

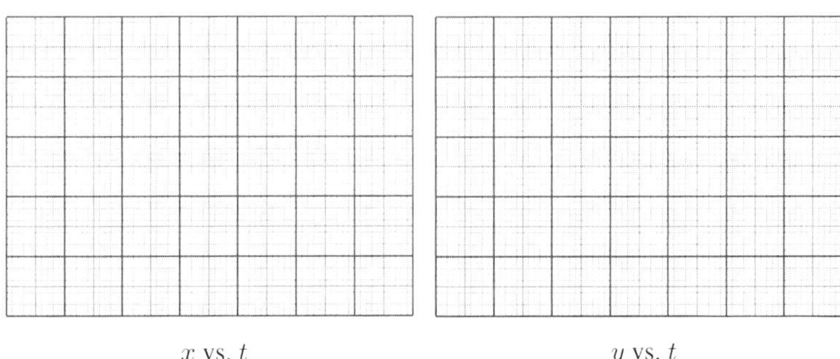

x vs. t $\quad$ y vs. t

3. Does either data set look like it can be modeled by a linear function $f(t) = mt + b$? If so, graphically estimate the value of m and b.

4. Compute the error between your model and the data. To do this, sum the absolute value of the difference between your model and the data at each recorded time. Can you change the value of m or b in your model to make the error smaller?

5. Repeat the above steps by plotting the changes in position versus time, Δx vs. t and Δy vs. t.

1.4 Function Composition Using Crackers and Cheese

Julie Barnes, Western Carolina University

Concepts Taught: function composition

Activity Overview

When students are introduced to function composition, it often has no meaning to them. Without any meaning, students may compose the wrong direction, turn composition into multiplication, or make any number of other mistakes. However, most students already have a very clear understanding of both the concept of a sandwich and how to make one. In this activity, sandwich making is broken down into two steps, each introduced as a function; students then use crackers and cheese to represent a variety of different compositions of the two functions. By changing the order of the sandwich-making steps, and relating this to function notation, students have an opportunity to make that connection between the notation for function composition and its meaning.

Supplies Needed (per group)	Class Time Required	Group Size
1 individual snack pack of crackers and cheese* Small paper plate	10 minutes	2-4 students

*The snack packs come in boxes and are convenient to use in class because they are easy to distribute and are the right size (roughly 5 or 6 crackers) for a group. It is also possible to use a large box of crackers and anything spreadable like cheese, jelly, or jam; for this to work, have a station where students can pick up the crackers, a plastic knife, and a spoonful of cheese, jelly, etc., on their plates for use during the activity.

Running the Activity

Provide each group with a paper plate, a snack pack of cheese and crackers, and a handout. Have students draw an "x" in the middle of the plate, as seen in Figure 1.5, and define the following two functions that will be used in this activity.

$C(x)$ is the function of placing a cracker on x.
$S(x)$ is the function of spreading one teaspoon of cheese on x.

It is helpful to have a brief discussion about the role of x as a placeholder and not just the mark on their plate. Ask questions such as, "What would C(chair) or C(hand) represent?"

Figure 1.5: Starting by drawing an "X" on a paper plate.

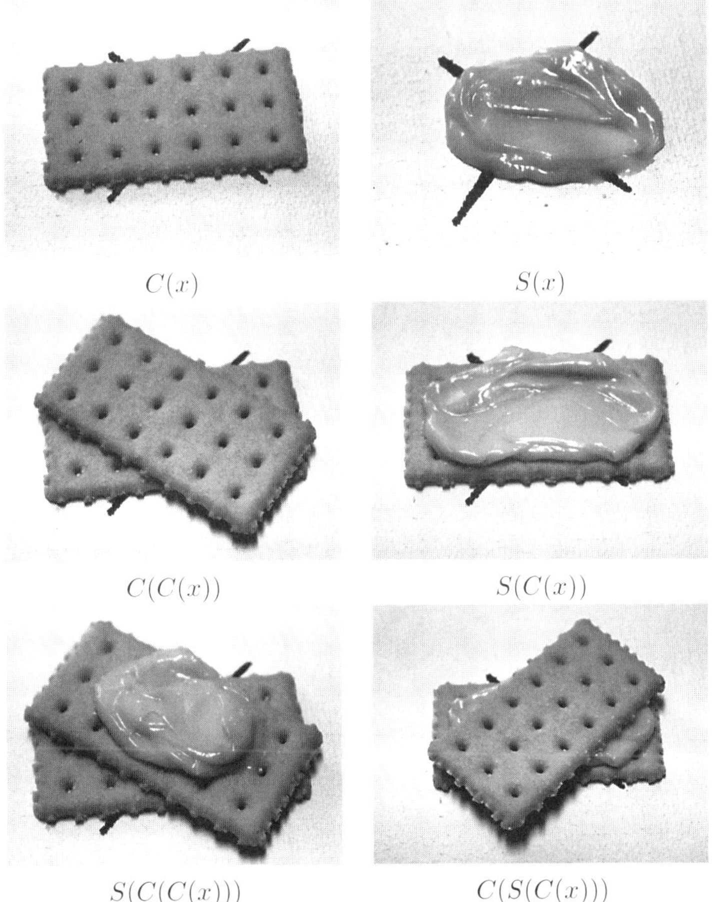

$C(x)$ $S(x)$

$C(C(x))$ $S(C(x))$

$S(C(C(x)))$ $C(S(C(x)))$

Figure 1.6: Photographs of constructions from the handout.

or "Would you want to create S(chair)? Why or why not?" Once everyone is comfortable with the notation, have them work through the problems on the handout. The handout asks students to explore a variety of possible compositions of $C(x)$ and $S(x)$ by creating each composition with their crackers and cheese. While students are working, circulate through the room to provide hints and answer questions. See Figure 1.6 for examples of the kinds of function compositions students will create.

Once the students have built their cracker creations, have a class discussion about the activity. What did they notice about the order of steps? How does this relate to the symbols describing what they did? How is composition different from basic addition? If they do not see a difference, ask them what $C(x) + S(x)$ would look like. It would have to involve two plates; one would have a cracker and one would have cheese spread on x. This is clearly different from what they just created.

Suggestions and Pitfalls

This is an easy activity to implement but it could generate crumbs; it is a good idea to bring paper towels or baby wipes. Also, be aware that students may have food allergies and some might not want to touch crackers or cheese for this reason. For a large class, you may want to do this as a class demonstration.

Function Composition using Crackers and Cheese – Class Handout

Before you begin, make sure that you have a paper plate and a snack pack of cheese and crackers. Draw an "x" in the center of your paper plate. We define the following two functions.

$C(x)$ is the function of placing a cracker on x.
$S(x)$ is the function of spreading one teaspoon of cheese on x.

1. Create $C(x)$. Describe what is on your plate.

2. Create $C(C(x))$. Describe what is on your plate. How does this differ from the previous problems?

3. Create $S(C(x))$. Describe what is on your plate.

4. Create $S(C(C(x)))$. Describe what is on your plate.

5. Make a cheese sandwich and place it on the "x" on your plate. What function symbols describe the sandwich you just created?

6. Would you want to create $C(S(x))$? Why or why not?

7. What is the difference between $C(S(x))$ and $S(C(x))$? Does order matter?

1.5 Walking Function Transformations

Julie Barnes, Western Carolina University
Kathy Jaqua, Western Carolina University

Concepts Taught: graphing function transformations, compositions of functions

Activity Overview

Once students have learned about functions and their related graphs, we often teach them how changes in the function translate into changes in the graph using the language of motion like *shifting*, *stretching*, *shrinking*, and *flipping*; yet we usually do not use the actual motions. In this activity, we give students an opportunity to become points on a Cartesian coordinate system and to walk through the motions dictated by transforming functions. In this way, students learn about shifting by shifting, stretching by stretching, and so on.

Supplies Needed	Class Time Required	Group Size
20 feet of adding machine paper* Tape Marker	50 minutes	3-4 students

*Painter's tape or string can be used instead of the adding machine paper.

Before class, rip the adding machine paper in half and tape it to the floor to represent an xy-coordinate system. Label units roughly one foot apart on the paper with a marker. If your classroom has moveable desks and a good amount of space, consider moving the desks to one side of the room to clear up some floor space.

Running the Activity

Provide each student with a copy of the handout and divide the class into groups of 3-4 students. Give each group an initial location with coordinates less than half of the largest x and y coordinates on your axes. Choose locations that are spread out on the coordinate system, making sure that one is on the x-axis and one is on the y-axis. Do Example 1 from Section 1 of the handout together. Then, have students complete Section 1 of the handout except for the final column. While they are working, walk around to answer questions and check their answers. If any groups finish early, have them help groups that are struggling.

Once all groups have completed Section 1, have one representative from each group stand at their given starting location and look at the first transformation of the handout. Have students *walk the transformation*, i.e., start at their given points and move to new locations as determined by the transformation. After walking the first transformation, have students describe what happened and fill in the last column on the chart in the handout. As you move on to the next transformation in Section 1, replace the volunteers so that by the end of the activity all students get a chance to participate; make sure the new walkers begin at their given starting points. See Figure 1.7 for photographs of where students would move under some of the transformations listed in the handout.

Once students have determined how to move under each type of transformation, challenge them to walk the transformations listed at the end of Section 1 without doing any computations.

Repeat this process for Section 2 of the handout.

At the end, ask students to explain or describe the difference between the transformations in Section 1 and Section 2 and challenge them to walk the transformations listed at the bottom which combine various transformations from both sections.

Suggestions and Pitfalls

Students tend to have more difficulty with transformations from Section 2 of the handout. For example, for $f(x-1)$, students often want to move to the left instead of the right. For $f(2x)$, they tend to want to stretch the function away from the y-axis instead of compressing it. Finally, a surprising thing may occur with $-f(x)$: students will sometimes stay in the same spot and spin 180 degrees. This is a good time to explain exactly what multiplying by a negative does.

This activity could also be done with more students walking if there is enough space. Just make room for them. For small classes, you can make the groups smaller or have students work individually.

For a more in-depth discussion, see [1] where this activity first appeared.

Reference

1. J. Barnes and K. Jaqua, Algebra aerobics, *Mathematics Teacher* 105 no. 2 (2011) 97–101.

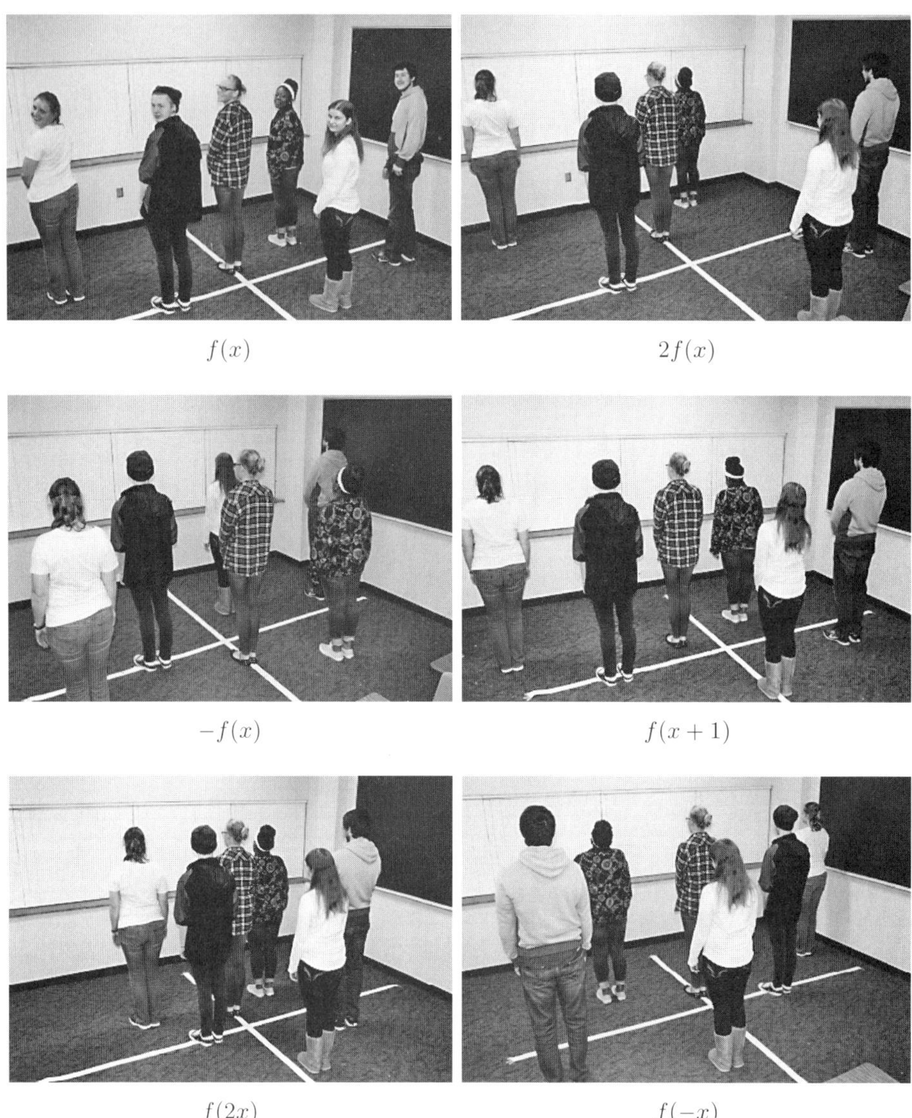

$f(x)$ $2f(x)$

$-f(x)$ $f(x+1)$

$f(2x)$ $f(-x)$

Figure 1.7: Some examples of students walking transformations from the handout. Note that the positive x-axis is to the right and the students are facing the positive y-axis. The starting points for $f(x)$ are $(-4, 2)$, $(-2, 0)$, $(0, 1)$, $(1, -2)$, $(2, 2)$, and $(4, -1)$.

Walking Function Transformations – Class Handout

My group's initial location is $(x, f(x)) = (\_\_\_\_\_\_\_\_, \_\_\_\_\_\_\_\_)$.

Section 1

In this section, fix your given x-coordinate and compute a new y-coordinate based on the transformation.

Example 1: Suppose the transformation is $f(x) + 1$ and your starting location is $(5, 6) = (x, f(x))$. Keep $x = 5$. Then, $y = f(x) + 1 = f(5) + 1 = 6 + 1 = 7$. The new location keeps $x = 5$, and now $y = 7$. Your new location is then $(5, 7)$.

Determine where your location moves under each of the transformations below. Fill in the first three columns.

Transformation	Fixed x Coordinate x	New y Coordinate y	New Location (x, y)	Description of Change
$f(x) + 1$				
$2f(x)$				
$\frac{1}{2}f(x)$				
$-f(x)$				

Stop here and wait for directions for whole class activity. You will be filling in the last column of the chart. Then some transformations you will be walking as a class are as follows.

$$f_1(x) = f(x) + 3, \quad f_2(x) = f(x) - 2, \quad f_3(x) = -2f(x), \quad f_4(x) = \frac{1}{2}f(x) - 2.$$

Section 2

In this section, fix your given y-coordinate and determine the new x-coordinate that would result in your fixed y coordinate after applying the function.

Suppose the transformation is $f(x+1)$ and your starting location is $(5,6) = (x, f(x))$. Keep $y = 6 = f(x+1) = f(5)$. Then, $x+1 = 5$, and $x = 4$. The new location keeps $y = 6$, and now $x = 4$. Your new location is then $(4,6)$.

Determine where your location moves under each of the transformations below. Fill in the first three columns.

Transformation	New x Coordinate x	Fixed y Coordinate y	New Location (x, y)	Description of Change
$f(x+1)$				
$f(2x)$				
$f(\frac{1}{2}x)$				
$f(-x)$				

Stop here and wait for directions for whole class activity. You will be filling in the last column of the chart. Then some transformations you will be walking as a class are as follows. $g_1 = f(x+3), g_2 = f(x-2), g_3 = f(-2x)$.

Final challenge: Without knowing your initial coordinates, how would you move when applying each of the following functions?

$$h_1(x) = f(x-1) + 2, \quad h_2(x) = 2f(x+1), \quad h_3(x) = f(2x) - 2.$$

1.6 Graphing Piecewise Functions with Feather Boas

Julie Barnes, Western Carolina University

Concepts Taught: piecewise functions, asymptotes, limits, graphs

Activity Overview

College students are fairly comfortable graphing standard linear and parabolic functions, if for no other reason than because they can punch some information into their calculator to obtain a good picture. However, the minute a function becomes piecewise, student confidence tends to dissipate.

In this activity, students work in small groups to graph two kinds of piecewise functions using feather boas. One of the functions is simply a combination of a linear piece and a quadratic piece. The other one is based on information provided about the limits of the function near asymptotes. It is easy to add more functions like these if you have time and your students could use more practice.

The beauty of using feather boas on the floor is that this type of group activity tends to catch the attention of students, making them want to participate. Since most students want to touch the feather boas, you typically do not have one student taking over and doing the activity alone. Also, the boa graphs are large enough that several students are able to work on them at the same time, and you are able to see how they are doing from a distance. In addition, the boas stay in place fairly well.

Supplies Needed (per group)	Class Time Required	Group Size
2 feather boas* 4 pieces of tape 10 feet of adding machine paper**	20-30 minutes	2-4 students

*Feather boas can be purchased at most craft stores. Alternatively, yarn or clothesline also work.

**Painter's tape can be used instead of the adding machine paper.

Running the Activity

At the beginning of class, provide each group with a handout, four pieces of tape, two feather boas, and ten feet of adding machine paper ripped into two five-feet sections. Have students find a place on the floor that is roughly 5 ft $\times$ 5 ft and tape the adding machine

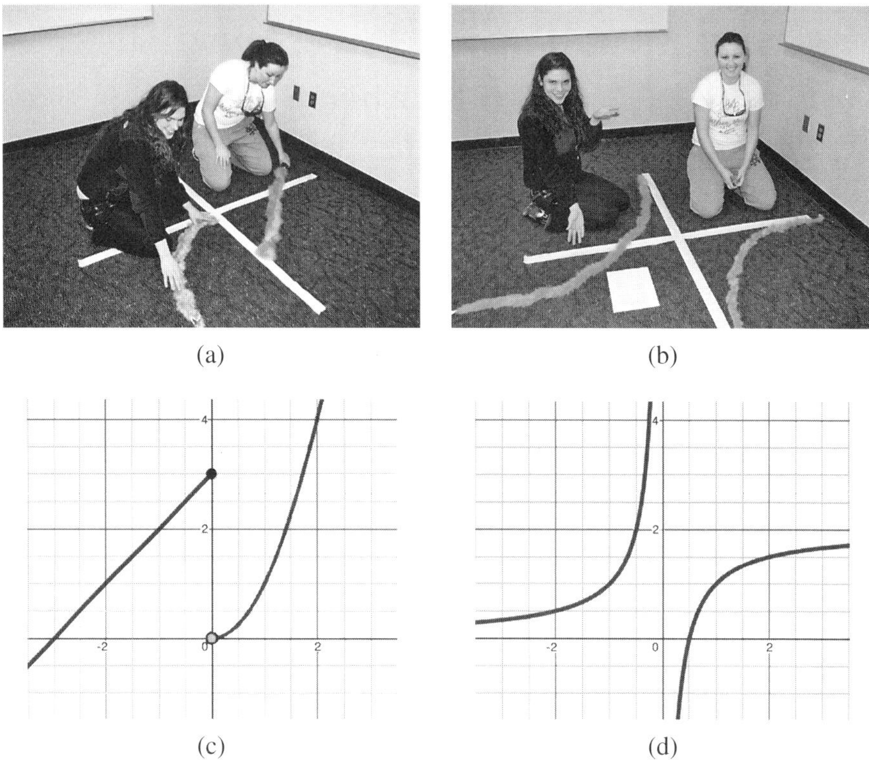

Figure 1.8: (a) Students working on each piece of the graph for $f(x)$. (b) Students displaying their graph for $g(x)$. (c) The graph for $f(x)$. (d) A possible graph for $g(x)$. Note that the graphs in (a) and (b) are oriented toward the students.

paper to the floor in such a way that it looks like a set of xy-axes as seen in Figures 1.8a and 1.8b.

Once their axes are ready, students should graph the problems stated on the handout. As they work, walk around the room to ask students questions about their graphs and have them explain their thinking. Be sure to have them indicate which is the x-axis and which is the y-axis, so everyone is reading it from the same direction. Examples of student work are shown in Figure 1.8, with Figures 1.8c and 1.8d included to show the graphs from the perspective of the students.

The goal here is to help students think through the meaning of a piecewise function and its corresponding graph; it is not just for students to generate perfect graphs, since that could be done more accurately with pencil and paper. In addition, since the boas are movable, students are able to try different ideas until they get the graph correct without needing to use an eraser. Also, capitalize on the fact that there are separate boas for each piece of the graph, as this allows students to think about the graph one piece at a time.

Suggestions and Pitfalls

When graphing $f(x)$, it is important to explain that the boas do not capture the open and closed circles (shown in Figure 1.8c). If desired, these features can be captured using additional props, such as canning rings for open circles and canning lids for closed circles.

Space is sometimes an issue. If your classroom is large enough with movable desks, have the students move the desks aside to work on the floor. Depending on logistics at your institution, you may be able to move into the hallway or any other nearby spaces like conference rooms or study areas. Alternatively, if the room has large tables, students can work on the tables instead of the floor.

Students often have a lot of questions. Therefore if you have a large class, it is best if you enlist a colleague or teaching assistant to help you on the day of the activity, as this is a great, non-threatening opportunity to provide some individual assistance while students work with the feather boas.

This activity can also be done in groups at their desks using Wikki Stix, Bendaroos, pipe cleaners, or even yarn. Wikki Stix and Bendaroos are available online and in some craft stores. These bendable sticks have the added feature that they adhere to the page but can also be removed much like sticky notes.

Graphing Piecewise Functions with Feather Boas – Class Handout

Use two pieces of adding machine paper that are each about five feet long in order to create an xy-coordinate system on the floor. Tape it in place. Then use two feather boas to create each of the graphs described below. When you finish creating each graph, sketch the graph and answer the questions.

1. Use feather boas to graph $f(x) = \begin{cases} x+3 & \text{if } x \leq 0 \\ x^2 & \text{if } x > 0. \end{cases}$

   Sketch a copy of your graph.

   (a) How does this graph relate to the individual graphs for $y = x + 3$ and $y = x^2$?

   (b) Why were you given two feather boas? Could this function have been graphed with only one boa? Why or why not?

   (c) Is your graph a function? How can you tell?

2. Use feather boas to graph a function that satisfies all of the following properties:

   $\lim_{x\to\infty} g(x) = 2$
   $\lim_{x\to 0^+} g(x) = -\infty$
   $\lim_{x\to 0^-} g(x) = \infty$
   $\lim_{x\to -\infty} g(x) = 0$

   Sketch a copy of your graph.

   (a) Are there any vertical asymptotes? If so, what are their equation(s)? How are they related to the limits?

   (b) Are there any horizontal asymptotes? If so, what are their equation(s)? How are they related to the limits?

   (c) Why were you given two feather boas? Could the function have been graphed with only one boa? Why or why not?

   (d) Is your graph a function? How can you tell?

1.7 Trigonometry Parameter Comparisons

Julie Barnes, Western Carolina University
Jessica Libertini, Virginia Military Institute

Concepts Taught: sine functions, amplitude, period, frequency, vertical shifts, phase shifts

Activity Overview

Students are typically able to graph trigonometric functions on a graphing calculator, and they can use the formulas for computing amplitude, period, vertical shifts, phase shifts, and frequency. However, they tend to have more difficulty explaining what the parameters of the functions mean and how they affect the corresponding graphs. In this activity, each student takes ownership of a particular sinusoidal graph and develops equation(s) that could generate his/her graph. In the process, students analyze the effect of the different parameters found in the the standard equation $y = a\sin(bx + c) + d$.

Supplies Needed	Class Time Required	Group Size
Prepared graph cards*	10-15 minutes	1-2 students

*Before the activity, photocopy the graphs from the table provided on pages 34 and 35 onto cardstock and use scissors to cut the graphs. You need one graph per group.

Running the Activity

This activity assumes that students have been introduced to the concepts and terminology of amplitude, period, vertical shifts, phase shifts, and frequency, as well as standard form for equations of sine graphs, but that they have had limited experience with these concepts.

Start by providing each group with one of the prepared graph cards and a handout. The handout then directs students as they work through four steps to create equations for their graphs. In the first step, students determine the amplitude, find other students whose graphs have the same amplitude, and then answer the questions from their handout concerning amplitude. Once everyone has completed this, students move to the second step where they determine the period of their graphs, find other students whose graphs have the same period, and answer the questions concerning the period of their graphs. Have students continue in this fashion as the handout guides them through a discussion about vertical shifts and phase shifts.

While students work, circulate around the room to help students find their groups and provide hints as needed. You may want to have some whole-class discussions to distill the main points after each step. Note that if you use the graphs provided, there should be four groups of four graphs in each of the first three steps. However, since phase shifts are not unique, there might be different sized groups of graphs in the last step. Students tend to have the most difficulty in determining other ways of computing phase shifts, so be prepared to provide hints like, "What would happen if you thought of it as a shift to the left?" or "What would happen if you add the period to your phase shift?"

When everyone has finished writing equations for their graphs, display them around the room, either by placing them on different desks, propping them on a chalkboard sill, or even taping them to the walls. Students can then walk around and compare the different equations with the graphs.

Suggestions and Pitfalls

There are 16 different graphs provided at the end of this section. If you have a smaller class, feel free to choose a subset of these graphs. For larger classes, students can work in groups or you can create more graphs for them to use.

If you want to do a similar activity for a wider collection of trigonometric functions like cosine curves or the negative sine function, feel free to add more graphs. If you do that, you may want to drop the phase shift portion of the activity, or simply remind your students that there is more than one trigonometric function that can be used to describe each graph.

On the top of the handout provided, the standard form for the sine function is written as $y = a\sin(bx + c) + d$. If you use a different standard form, you may want to make changes accordingly.

The equations used to generate the graphs on pages 34-35 are listed in Table 1.3. Note that by changing phase shifts, there is more than one possible equation for each of the graphs.

Table 1.3: Equations used to generate graphs found on pages 34 and 35.

A.	$y = \sin(x)$	B.	$y = \sin(2x - \pi) + 1$	C.	$y = 2\sin(4x)$
D.	$y = 2\sin\left(x - \frac{\pi}{2}\right) + 1$	E.	$y = \frac{1}{2}\sin\left(\frac{1}{2}x - \frac{\pi}{4}\right)$	F.	$y = \frac{1}{2}\sin(4x - \pi) + 1$
G.	$y = \frac{3}{2}\sin(2x - \pi)$	H.	$y = \frac{3}{2}\sin\left(\frac{1}{2}x\right) + 1$	I.	$y = \sin\left(\frac{1}{2}x + \frac{\pi}{4}\right) + \frac{1}{2}$
J.	$y = \sin(4x) - 1$	K.	$y = 2\sin(2x + \pi) + \frac{1}{2}$	L.	$y = 2\sin\left(\frac{1}{2}x\right) - 1$
M.	$y = \frac{1}{2}\sin\left(x + \frac{\pi}{2}\right) + \frac{1}{2}$	N.	$y = \frac{1}{2}\sin\left(2x - \frac{\pi}{2}\right) - 1$	O.	$y = \frac{3}{2}\sin(4x - \pi) + \frac{1}{2}$
P.	$y = \frac{3}{2}\sin\left(x + \frac{\pi}{2}\right) - 1$				

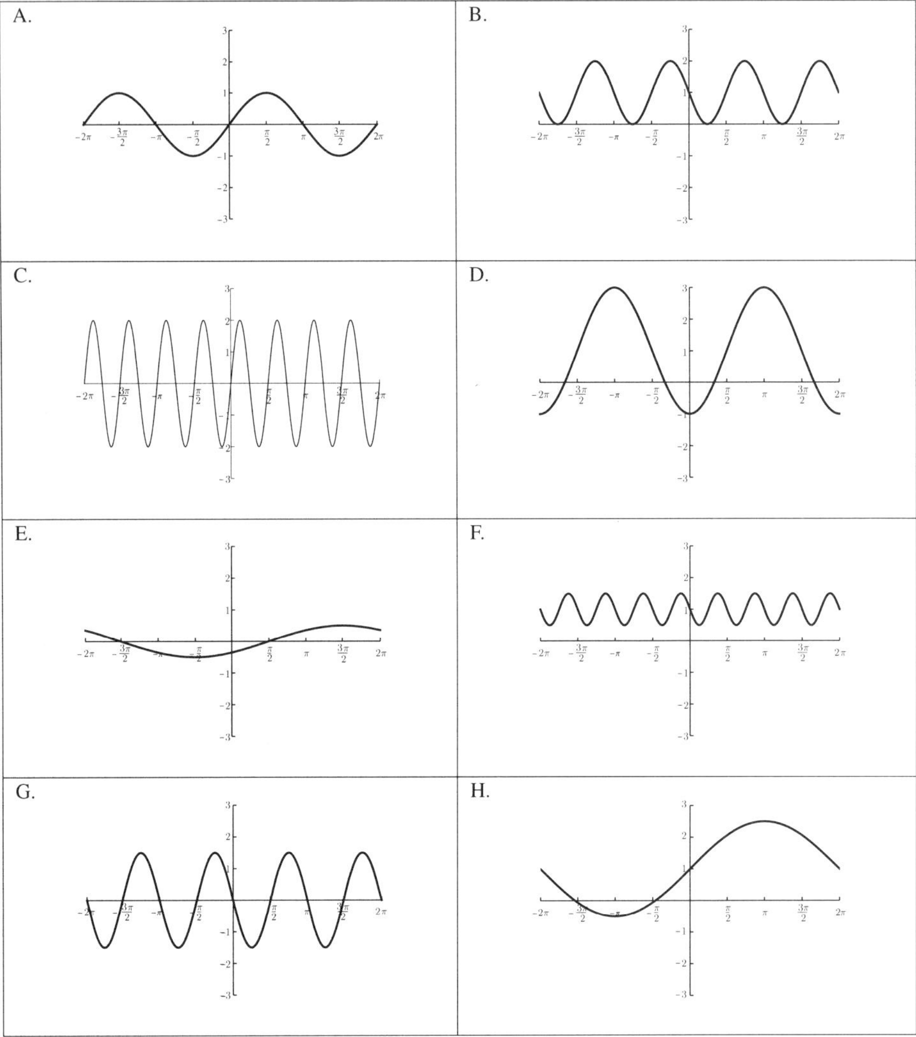
A.
B.
C.
D.
E.
F.
G.
H.

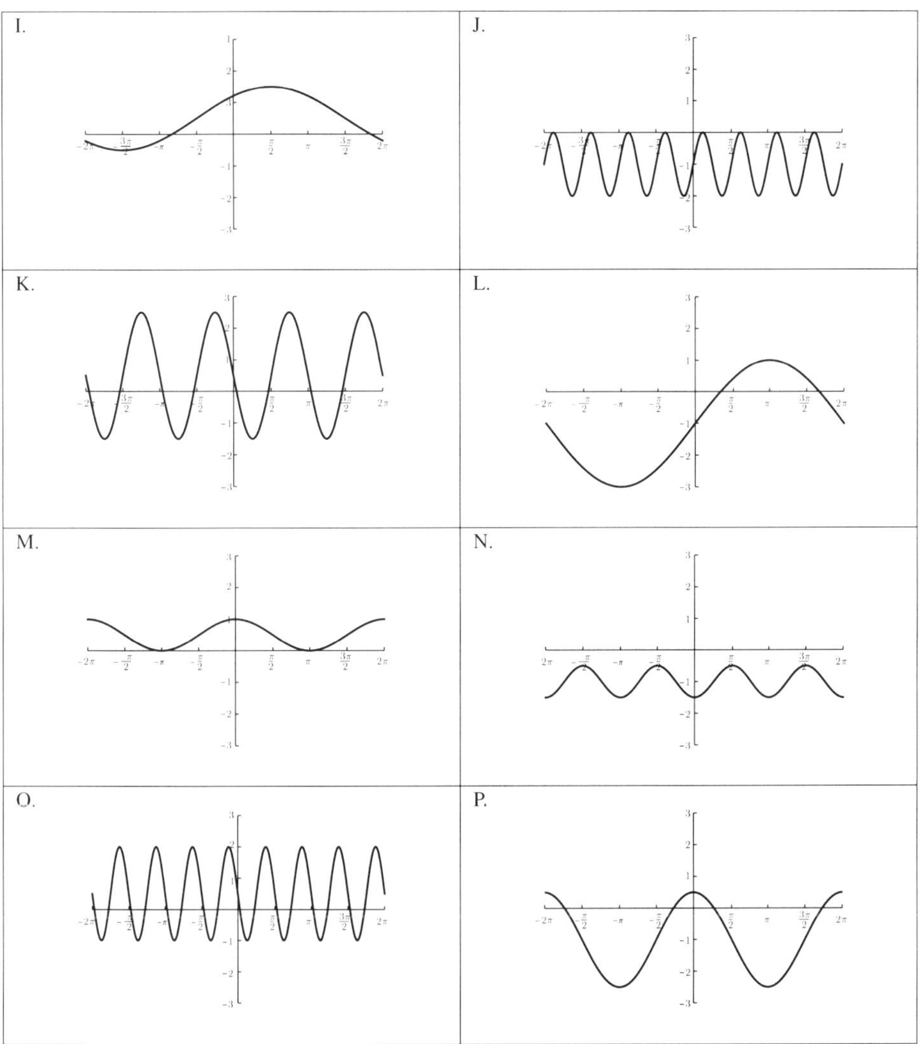
I.
J.
K.
L.
M.
N.
O.
P.

Trigonometry Parameter Comparisons – Class Handout

In this activity, you will go through several steps to develop an equation for your graph. All functions shown could be written in the form $y = a\sin(bx + c) + d$.

Amplitude

1. Determine the amplitude of your graph. Then, find all of your classmates who have graphs of functions with the same amplitude.

2. Describe the similarities in the collection of graphs that have the same amplitude as your graph.

3. For each graph in your group, determine the maximum value (height), the minimum value, and the value of the maximum value minus the minimum value. How is this calculation related to the amplitude?

4. Use the amplitude to determine the a portion of your equation, and then wait for your instructor to ask you to move to the next step.

Period

1. Determine the period of your graph. Then, find all of your classmates who have graphs of functions with the same period.

2. Describe the similarities in the collection of graphs that have the same period as your graph.

3. For each graph in your group, calculate the frequency. How does this number compare to the period? What is the difference between period and frequency graphically?

4. Use the period to determine the b portion of your equation, and then wait for your instructor to ask you to move to the next step.

Vertical Shift

1. Determine the vertical shift of your graph, keeping in mind that there might not be a vertical shift (i.e., the vertical shift is zero). Then, find all of your classmates who have graphs of functions with the same vertical shift.

2. Describe the similarities in the collection of graphs in your group that have the same vertical shift.

3. Use the vertical shift to determine the d value in your equation, and then wait for your instructor to ask you to move to the next step.

Phase Shifts

1. Determine a phase shift for your graph, keeping in mind that there might not be a phase shift (i.e., the phase shift is 0). Then, find all of your classmates who have graphs of functions with the same phase shift. If you are having trouble finding anyone, try computing another value for the phase shift.

2. Describe the similarities in the collection of graphs in your group that have the same phase shifts.

3. As a group, determine two other phase shifts that would generate the same graph. How could you describe an infinite number of other phase shifts that could also be used to generate the same graph?

4. Use one of your phase shifts to determine a value for c in your equation.

Part II

Main Courses (Calculus)

Chapter 2

Differential Calculus

2.1 Adding Movement to Velocity Explorations with Ziplines

Lisa Driskell, Colorado Mesa University
Audrey Malagon, Virginia Wesleyan College

Concepts Taught: average velocity, instantaneous velocity, motivation of limits

Activity Overview

Students entering calculus often understand the computations for average velocity but have not considered the issues that arise when using their formulas to compute instantaneous velocity. In this activity, students construct a zipline and record times and distances to measure average velocity over smaller and smaller intervals. This activity covers average velocity calculations, the concept of instantaneous velocity, and the relationship between the two, setting up a framework for the study of limits later in the course.

Supplies Needed (per group)	Class Time Required	Group Size
Zipliner* Smooth gift wrapping ribbon at least 6-10 feet each Tape measure Tape Marker Stopwatch or smart phone app	30-50 minutes Longer time allows for more explanation but is not required	3-4 students

*A *zipliner* is any weighted object, such as a balloon with a marble in it, on a key ring or binder clip. See Figure 2.1a.

Running the Activity

After an introductory discussion about the activity, provide each group with the supplies needed and a handout. Part I of the handout guides students as each group builds and tests a zipline. To build the zipline, students need to pass the ribbon through the key ring or binder clip attached to the zipliner (Figure 2.1a), and with tape secure the ends of the ribbon to walls, desks, chairs, or the floor (Figure 2.1b). The ribbon zipline should be taut with no twists and should be angled so the zipliner takes at least two seconds to smoothly travel the entire length. Acceptable angles typically range between 25 and 45 degrees. Building

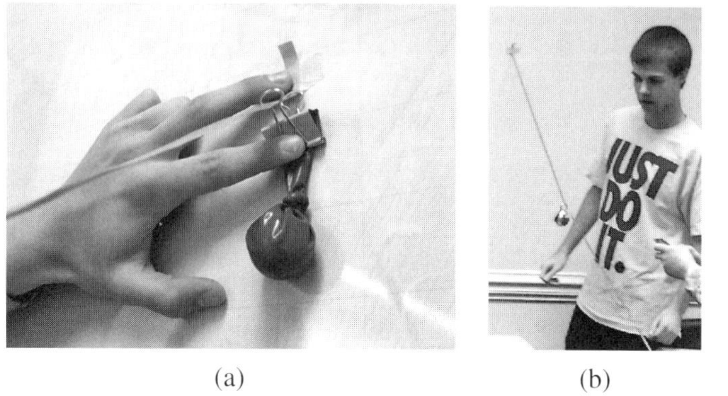

(a) (b)

Figure 2.1: (a) A zipliner on a piece of ribbon. (b) Riding on the zipline.

and testing should take less than five minutes. Once the ziplines are built, have students continue working through the handout while you walk around providing assistance.

In Part II, students explore the information they are able to determine with just the materials provided. Ambitious groups may calculate angles and slopes, but almost all record time and distance. Keep this exploration part to no more than five minutes. You may want to invite the groups to share their ideas with the class. The handout continues by asking students to work with average velocity. Students first calculate and record the average velocity of their zipliner for the entire trip. This is done by having the students use a stopwatch or stopwatch app to time the zipliner as it travels from start to finish.

Next, the handout asks students to mark the halfway point on the zipline. Students record the average velocity for the top half of the trip as well as the bottom half of the trip. Be sure to remind students to always start their zipliners from the top, regardless of what portion of the zipliners' movement is being recorded. Once students have computed these average velocities, they compare these values along with their initial recording of average velocity. Students are often amazed that the values are different. The handout asks students to discuss the results with their group members. Because the results are surprising to many students, you should have the groups share their findings with the class and then facilitate a discussion about speed versus average speed, possibly using a student's drive to school that morning as an example. You could ask students how the speedometer knows how fast the car is going at any given moment to start them thinking about what instantaneous velocity means.

Part III of the handout guides the students toward thinking about limits and instantaneous velocity by measuring additional average velocities. These averages are computed over shorter and shorter intervals that surround the halfway mark.

Part IV asks students to consider whether or not it is possible to know exactly how fast the zipliner is traveling the moment it passes the halfway mark. If they think it is possible, they are asked to explain how they might estimate the instantaneous velocity. As you assess their individual writing assignments, note that a strong case can be made for either a yes or a no answer.

Suggestions and Pitfalls

Because this activity highlights the differences between average and instantaneous velocity, it is crucial to be very precise with this language when talking to the class. Always explicitly state "average velocity" or "instantaneous velocity" during class discussions, and ask students to specify which they mean if they use the term "velocity" alone. Students at this stage are not always able to infer the type of velocity from the context, or worse, they interpret our imprecision as meaning that all velocities are the same.

You may also need to remind students several times to always start their zipliner from the top of the zipline and to allow it to travel to the end of the zipline, no matter which interval they are measuring. Many students will start the zipliner at the halfway mark to measure the bottom half, if not reminded.

This activity provides a physical and visual experience that illustrates a need for limits that can be referenced throughout the semester. The activity relates well to the introduction of derivatives while the concept of using increasingly smaller intervals of time relates well to topics like net change.

For a more in-depth discussion, see [1] where this activity first appeared.

Reference

1. L. Driskell and A. Malagon, Keychain ziplines: A practical way to study velocity in the calculus classroom, *PRIMUS* 23 no. 7 (2013) 590–597.

Adding Movement to Velocity Explorations with Ziplines – Class Handout

Part I: Build

1. Obtain a zipliner, a piece of ribbon, a tape measure, tape, a marker, and a stopwatch.

2. Place the zipliner on the ribbon and tape the ends of the ribbon to walls, desks, etc., to make a zipline. Be sure the line is taut and at an angle.

3. Test the zipline. The zipliner should travel smoothly from top to bottom without any pauses or bouncing. Use a stopwatch or stopwatch app to time the zipliner. Verify that it takes at least two seconds for the zipliner to travel the full length of the ribbon. Adjust the zipline, if needed, and retest.

Part II: Explore

1. Using only the tools you've used thus far, what information can you determine? Discuss this with your group members and write down some ideas and observations.

2. Choose a member of the group to time the zipliner on several trips until the timing is consistent.

   i. Record the length of the zipline. Use appropriate units.
   ii. Record the time it takes the zipliner to travel the zipline. You may wish to run the zipliner a few times and use an average of those times as your time. Use appropriate units.
   iii. Write the formula for average velocity, and find the average velocity of the zipliner for the entire trip. What are the units?

3. Find and mark the point on the ribbon that is half of the distance from the bottom of the zipline. When calculating average velocities, **every zipline run should travel the entire length of the ribbon**.

   i. Find the average velocity of the zipliner while traveling the top half of the zipline.
   ii. Find the average velocity of the zipliner while traveling the bottom half of the zipline.
   iii. Were the average velocities the same? How do the average velocities compare to the average velocity of the trip from top to bottom? Can you give an explanation for the differences?

Part III: Discover

We noticed in Part II that the average velocity for the top half and the bottom half of the zipline were different. Now we will investigage average velocities through the middle of the zipline. Remember: **every zipline run should travel the entire length of the ribbon.** You also may wish to run the zipliner a few times and use an average of those times for each item below.

1. Find the average velocity of the zipliner as it travels through the middle half of the zipline. To do this, first mark the points a fourth of the distance from the top and a fourth of the distance from the bottom of the ribbon and record the distance between the two marks. Then find the average velocity of the zipliner as it travels between those marks.

2. Find the average velocity of the zipliner as it travels through the middle third of the zipline. To do this, mark the points a third of the distance from the top and from the bottom of the ribbon and record the distance between the two marks. Then find the average velocity of the zipliner as it travels between those marks.

3. (Time permitting) Choose another interval shorter than the middle third that still includes the halfway mark. Mark a point on the ribbon above the halfway mark and one below the halfway mark. Record the distance between the new marks. Find the average velocity of the zipliner as it travels between the two marks.

PART IV: Inquire

1. We have measured average velocities by calculating distance traveled and the time it took to travel that distance. Is it possible to know exactly how fast the zipliner is traveling the moment it passes the halfway mark? If so, describe how you would determine this instantaneous velocity. If not, explain why not and explain how you might estimate the instantaneous velocity.

2. Writing Assignment: Prepare a brief write-up discussing today's activity. Summarize your group's findings and discuss the question in Part IV. Please type your entry.

2.2 Creating Limit Windows with Index Cards

Julie Barnes, Western Carolina University
Jessica Libertini, Virginia Military Institute

Concepts Taught: limits, one-sided limits, two-sided limits

Activity Overview

When students enter a calculus class, they are often most comfortable doing straightforward computations and less comfortable with conceptual topics. One of the more difficult early concepts is the notion of approaching a point without necessarily getting there and how that idea relates to one-sided and two-sided limits. Graphs of functions allow students to visualize limits, but some students may have difficulty focusing on the correct portion of the graph and thus read it incorrectly. In this activity, students use index cards to cover up different portions of a graph to demonstrate the idea of one-sided and two-sided limits while determining the value of the limit.

Supplies Needed (per group)	Class Time Required	Group Size
2 index cards	15 minutes	2-3 students

Running the Activity

Provide each group of students with a handout and two index cards. Then, demonstrate to the class how to use index cards to do the first two examples on the handout; this works best if you use a document presenter to project your demonstration.

For Example 1, demonstrate $\lim_{x\to 0^+} f(x)$. Start by covering up the left side of the graph with one of the index cards, reminding students that we are not using that portion of the graph. Then move the right index card towards 0 while students do the same thing with their cards. Ask them what the height of the function near the edge of the right card is approaching as you slide it towards 0. See Figures 2.2a, 2.2b, and 2.2c for three snapshots of what happens in Example 1 as you represent $\lim_{x\to 0^+} f(x)$. Similarly, demonstrate how to find $\lim_{x\to 0^-} f(x)$.

For Example 2, use both index cards to demonstrate $\lim_{x\to 0} f(x)$; that is, lay the index cards down on each side of the graph and slowly move the cards together, meeting at $x = 0$. As you slide the cards together, have students do the same thing on their handout and ask them what the height of the function is on the edges of the index cards as you move them inward. See Figures 2.2d, 2.2e, and 2.2f for three snapshots of what happens in Example 1 as you slide the cards together.

Have the students work through the rest of the handout as you walk around to answer questions and provide assistance. Be sure to observe their process as they use the note cards in determining their limits.

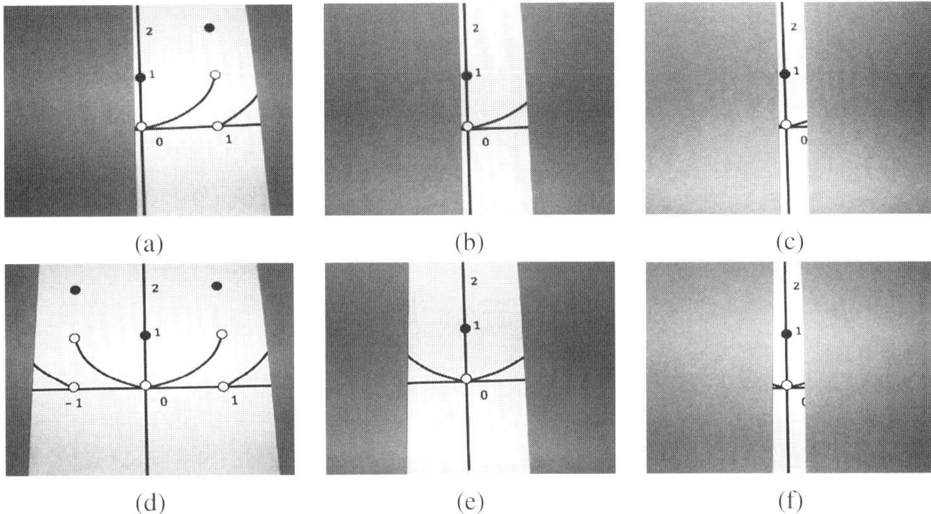

Figure 2.2: (a)-(c) Three different stages as an index card is moved inward to demonstrate $\lim_{x\to 0^+} f(x)$. Note that one card covers the left side of the graph, indicating that the portion of the graph to the left of 0 has no bearing on the limit from the right. (d)-(f) Three different stages as index cards are moved inward to demonstrate $\lim_{x\to 0} f(x)$. Note that as we move from (d) to (f), the cards are converging to the limit of $f(x)$ at 0.

Suggestions and Pitfalls

The cards help make sure students are looking at the correct region, but they still may struggle with problems involving open and closed circles on piecewise functions. It is helpful to ask questions of students while they work to make sure they grasp the relationship between limits and piecewise functions.

Creating Limit Windows with Index Cards – Class Handout

Use the function $f(x)$ to answer the problems below. For each problem, answer the question, and also record notes about any observations you make while using the index cards to evaluate various limits.

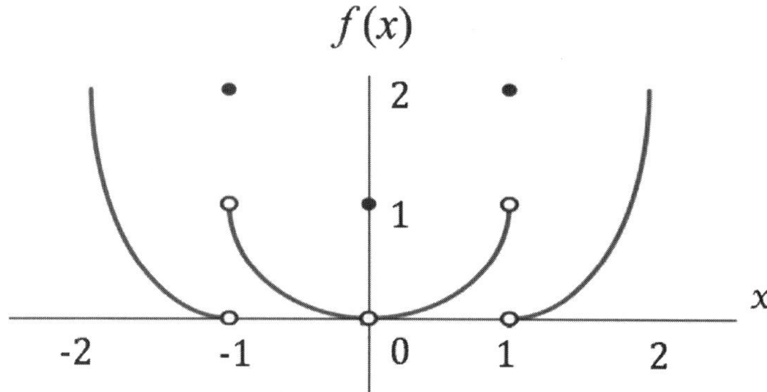

Example 1: Use the index cards to evaluate the $\lim_{x\to 0^+} f(x)$ and $\lim_{x\to 0^-} f(x)$.

Example 2: Use the index cards to evaluate the $\lim_{x\to 0} f(x)$.

1. What is the value of $f(0)$? Is this the same or different from $\lim_{x\to 0} f(x)$? What is different about how we use the cards to evaluate the limit in Example 2 and how we evaluate the function at 0?

2. How is the process of finding the limit in Example 1 different from the process found in Example 2? How do the index cards show the difference?

3. Use cards to evaluate $\lim_{x\to 1^+} f(x)$.

4. Use cards to evaluate $\lim_{x\to 1^-} f(x)$.

5. Use cards to evaluate $\lim_{x\to 1} f(x)$.

6. What is the same or different about using the cards to evaluate $\lim_{x\to 1} f(x)$ and $\lim_{x\to 0} f(x)$?

7. Use cards to evaluate $\lim_{x\to -1^+} f(x)$.

8. Use cards to evaluate $\lim_{x\to -1^-} f(x)$.

9. Use cards to evaluate $\lim_{x\to -1} f(x)$.

10. Use cards to evaluate $\lim_{x\to \frac{1}{2}} f(x)$. You will need to estimate the approximate location of 1/2 on the graph as well as estimate the value of the limit. How is this limit the same or different from the one in Example 1?

2.3 Extreme Values with Pipe Cleaners

Karen Bliss, Virginia Military Institute
Jessica Libertini, Virginia Military Institute

Concepts Taught: extreme values

Activity Overview

For most instructors, it makes intuitive sense that for a continuous function on a closed interval, the global extrema must occur at critical points or endpoints. Calculus students, however, often struggle to internalize this concept, perhaps because when this appears as a theorem, the notation and language are abstract. In this activity, students attempt to use pipe cleaners to design the graphs of functions whose global extrema meet certain properties—some possible, some impossible. They are led in a very natural way to discover the extreme value theorem for themselves.

Supplies Needed (per group)	Class Time Required	Group Size
8 pipe cleaners* Blank paper	30 minutes	2 students

Anything flexible and at least 6″ long could be used, such as Benderoos®, Wikki Stix® or cut yarn.

Running the Activity

As this is an exploratory activity in which students will discover the extreme value theorem, it is best if you run this activity before teaching the theorem. Have students sit in pairs at their workspaces, and pass out the pipe cleaners and handouts.

As students work to create functions that satisfy the criteria given in the questions on the handout, circulate in the room, asking students about their thoughts and providing assistance and guidance as needed. If you find groups with conflicting answers, it is useful to ask them to compare their answers and come to consensus.

After most groups have completed the handout, lead the whole class in a discussion, asking the students to identify the commonalities among the exercises. Eventually, you will want to get students to articulate the ideas of the extreme value theorem, at which point you can congratulate them for being amazing mathematicians who have discovered a very important theorem.

Suggestions and Pitfalls

Students should be able to create satisfactory functions for Problems 1, 2, 4, and 5, unlike those in Problems 3 and 6 which are not possible. Problems 4, 5, and 6 really drive home the extreme value theorem, so as students are working, make sure they have thoughtfully designed those graphs.

This activity is most powerful when students are given time to actually come up with the extreme value theorem (or something very close to it) on their own. You can play the role of recorder, writing their conjectures on the board and using phrases like, "I'm hearing this group say that if the global max isn't at an endpoint, then it must be at one of the local maxima. Do we all agree with that?"

Extreme Values with Pipe Cleaners – Class Handout

Below are several descriptions of functions. For each description, use a pipe cleaner to try to make the graph of a **continuous** function fitting the description. Call the x-value of the left end of your pipe cleaner a and the x-value at the right end b.

Note: Some are possible to create, but others may be impossible. For those that are possible, make a small sketch of the graph of your function on your paper in the space provided. For those that are impossible, give a brief explanation about why you found it impossible.

1. A function whose global maximum over $[a, b]$ is at $x = b$, and whose global minimum over $[a, b]$ is at $x = a$.

2. A function whose global maximum over $[a, b]$ is at a critical point in (a, b), and whose global minimum over $[a, b]$ is at one of the two endpoints, either a or b.

3. A function whose global maximum over $[a, b]$ is at a critical point in (a, b), and whose global minimum over $[a, b]$ is at an inflection point in (a, b).

4. A function whose global extrema (maximum and minimum) over $[a, b]$ are not located at endpoints.

5. A function whose global extrema over $[a, b]$ are not located at critical points of the function.

6. A function whose global maximum over $[a, b]$ is at an endpoint, and whose global minimum over $[a, b]$ is neither at an endpoint nor a critical point.

7. Based on the observations above, what can we say about where global extrema can occur for a function defined on a closed interval $[a, b]$?

2.4 Graphing Puzzles Using the First and Second Derivatives

Karen Bliss, Virginia Military Institute

Teena Carroll, Emory and Henry College

Jessica Libertini, Virginia Military Institute

Concepts Taught: sketching functions, first derivative, second derivative, continuity

Activity Overview

When students are learning what the first and second derivatives tell them about the graph of a function, they often get overwhelmed by the amount of information. While they typically understand the relationships one by one (for example, it's not hard for a student to be convinced that if $f'(x) > 0$, then $f(x)$ is increasing), they often stumble when the time comes to consolidate all of their knowledge.

In this activity, students reinforce the concepts they have already learned by practicing translating between the description of a function in words, the description in mathematical notation, and the actual shape of the function. Working in small groups, they fill out a table for each of three graphs and use puzzle pieces to assemble the graph of a continuous function that satisfies the given conditions.

Supplies Needed	Class Time Required	Group Size
Prepared sticky notes*	15-20 minutes	2-3 students

*Use a thick, black marker to draw on square sticky notes to create curved and linear puzzle pieces as in Figure 2.3. Alternatively, you can give the students pieces of paper that have all the required puzzle pieces on them, and then hand out scissors and ask them to cut out the pieces. Each group will need several copies of each piece in order to assemble puzzles. For the puzzles in the handout, each group will need two curvy pieces and three linear pieces.

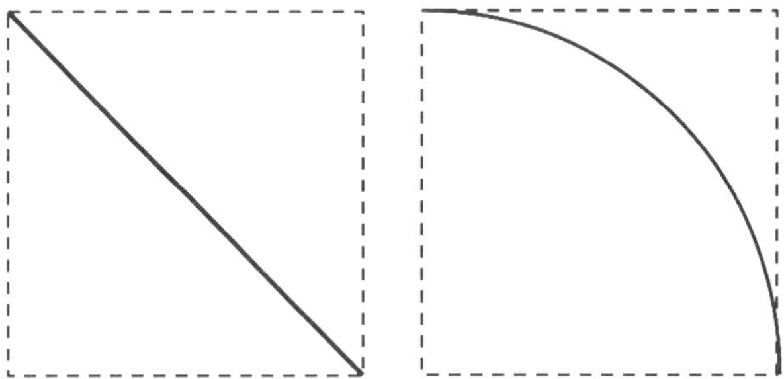

Figure 2.3: Sample linear and curved puzzle pieces.

Running the Activity

This activity is best used to reinforce the idea that information about the first and second derivatives can be used to learn about the shape of the original function. As such, it is helpful to start class with a brief review of these concepts and then begin the activity.

Provide each group a copy of the handout and a set of puzzle pieces or a paper with the pieces and scissors. The handout includes three columns, each describing the shape of a region of a function in a different way: verbally, symbolically, and graphically. For each of these regions, students should use the existing information from a column to fill in the other two columns. Each group works together to fill in the empty boxes on the handout. Once the tables on the handout are completed, students take their puzzle pieces and assemble the graph of each function, as shown in Figure 2.4.

As the students work, walk around the room to see if students are on track. Provide positive encouragement when appropriate; if a group is making errors, ask them to explain their thinking and prod them to identify and correct their own mistakes. As the groups create their graphs, verify their work, so that they don't potentially graph all of them incorrectly before you are able to intervene.

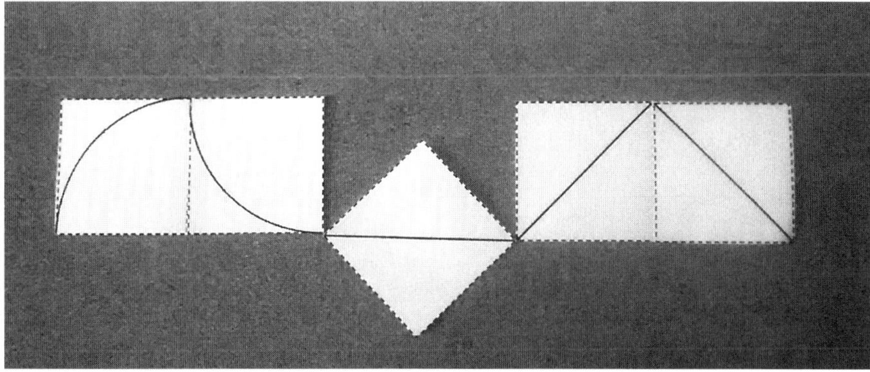

Figure 2.4: Example of an assembled puzzle. This is Graph C from the handout.

Suggestions and Pitfalls

It is not advisable to have students draw their own puzzle pieces, as it is too easy to draw puzzle pieces that make it impossible to create the functions.

In filling out the tables, students should be encouraged to share writing duty, so that each of them remains engaged in the activity and has a turn translating between the descriptions, mathematical notation, and the shape of the graph of the function.

When the time comes to assemble the graph, students may struggle with the idea of rotating the puzzle pieces. For example, a student may view the curved puzzle piece (as in Figure 2.3) as being a segment that is concave down and decreasing. They may need some prompting to see that if they rotate the piece 90° to the right, then it is concave up and increasing, and that, in fact, all four possible curved shapes can be created by rotating a single curved puzzle piece.

Some students also try to line the puzzle pieces up horizontally. You may need to remind them that the puzzle pieces need to be pushed up or down in order to align the endpoints and assemble the graph of a *continuous* function.

Instructors can extend this activity by asking students to identify points on the graphs of their continuous functions where there are local extrema, inflection points, or points where the function is nondifferentiable.

Graphing Puzzles Using the First and Second Derivatives – Class Handout

Use the information on each line to fill in the remaining columns. A couple of examples are provided for you.

GRAPH A

Interval	Description in words	Description in symbols	Sketch
1	concave up and increasing		
2	concave down and increasing	$f''(x) < 0$ and $f'(x) > 0$	
3	zero concavity and decreasing		
4	zero slope		
5	zero concavity and increasing		

GRAPH B

Interval	Description in words	Description in symbols	Sketch
1		$f''(x) = 0$ and $f'(x) > 0$	
2		$f''(x) = 0$and $f'(x) < 0$	
3		$f''(x) > 0$ and $f'(x) < 0$	
4	concave down & decreasing	$f''(x) < 0$ and $f'(x) < 0$	
5		$f'(x) - 0$	

GRAPH C

Interval	Description in words	Description in symbols	Sketch
1	concave down and increasing		
2			
3		$f'(x) = 0$	
4		$f''(x) = 0$ and $f'(x) > 0$	
5	decreasing and zero concavity		

Now that you've completed the tables, assemble the graphs of each of the **continuous** functions described. After completing each graph, verify your answer with your instructor.

2.5 Graphing Functions from Derivative Information Using Bendable Sticks

Julie Barnes, Western Carolina University

Concepts Taught: derivatives, critical points, increasing and decreasing, inflection points, concavity, graphs

Activity Overview

With the amount of technology available, students often expect calculators or computers to do all their graphing for them. However, this could keep them from making valuable connections between the derivatives of a function and the shape of that function's graph. In this activity, students use information about derivatives to graph functions without being provided with equations; this forces them to consider those connections between derivatives and graphs. Students are also able to experience the function properties of the first and second derivatives by physically creating a curve instead of just sketching it on paper.

Supplies Needed (per group)	Class Time Required	Group Size
3 bendable sticks, preferably in different colors	20-30 minutes	2-3 students

*Two brands of easy-to-use bendable sticks that work well are Wikki Stix® and Bendaroos® which can be found online and sometimes in craft stores. An added bonus to using these is that they will adhere to paper and yet can be removed much like sticky notes can be removed; therefore, graphs will stay in place while students are working. Pipe cleaners can also be used, but they don't stick to the paper nicely.

Running the Activity

Have each group of students begin by sketching a coordinate system on their paper for $-2 < x < 2$ and $-2 < y < 2$, with roughly $1''$ per unit. The only point that absolutely needs to be marked on the graph is $(1, 0)$. Provide each group with three different colored bendable sticks and a handout. Have students work through the handout in groups.

In the first problem, students are given a list of conditions about the first and second derivative of a function h and are asked to use a bendable stick to create a graph that meets the given criteria as seen in Figures 2.5a and 2.5d. Then they need to use a second bendable stick to graph the first derivative of h (Figures 2.5b and 2.5e) and a third to graph the second derivative of h (Figures 2.5c and 2.5f). Since the sticks can be moved and bent, students are able to work with different criteria in any order, possibly bending the stick enough to satisfy increasing and decreasing features and then bending further to obtain the correct

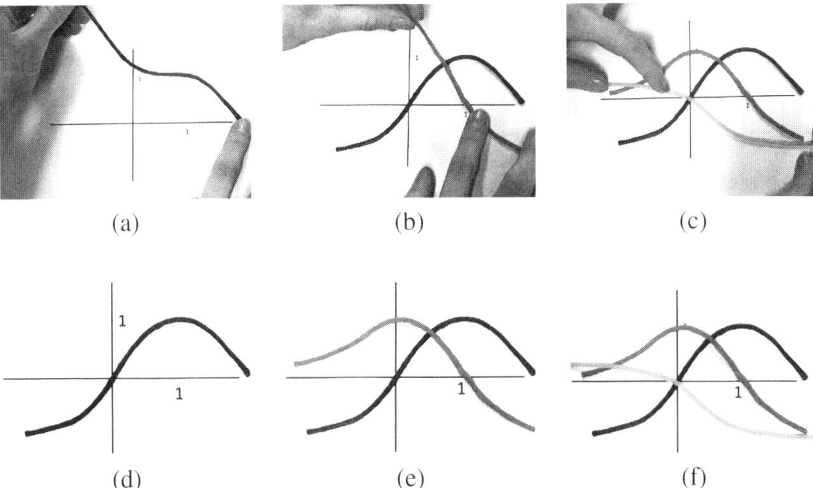

Figure 2.5: (a)-(c) Students using Wikki Stix to create graphs for h, h', and h'' as described in Problem 1 of the handout. (d) A possible graph for h. (e) A graph of h' with h. (f) A graph of h'' along with h' and h.

concavity. This means students can keep modifying their work until it meets all criteria, and they do not need to erase. Once students have a function and its derivatives graphed, they are asked to compare the derivatives to the information used to create the function in the first place.

The second problem asks students to graph any function of their choosing. Sometimes this catches them off guard because they are used to being told exactly what to graph instead of just being told to make sure it passes the vertical line test. Once they have a graph, they are asked to use a second bendable stick to graph a new function that has the same derivative. For students who are struggling, you can mention that the easiest way to do this is to bend the second stick on top of the first one, and then pick it up and move it up or down the y-axis. Make sure students understand why this works.

While the students work, circle around the room to answer questions and check their work. Note that because of the nature of the Wikki Stix, it is easy do things like start where the derivative is zero, and then bend the curve accordingly.

Suggestions and Pitfalls

This activity also works well with feather boas. In Figure 2.6 we see students working on Problem 2 from the handout. Feather boas take a little more setup time and space, as seen in Activity 1.6, Graphing Piecewise Functions with Feather Boas, but students tend to work more as a team with the larger objects.

Figure 2.6: Students using feather boas to graph two functions that have the same derivative, as in Problem 2 from the handout.

Graphing Functions from Derivative Information – Class Handout

Make sure you have three different colored bendable sticks. Sketch an xy-coordinate system your on paper for $-2 < x < 2$ and $-2 < y < 2$, with roughly $1''$ per unit, to be used throughout this activity.

1. Use bendable sticks to graph a function, $h(x)$ on your xy-coordinate system, that satisfies all of the following properties.

   $h(x)$ is continuous.
   $h''(0) = 0, h'(1) = 0$
   $h'(x) > 0$ when $x < 1$
   $h'(x) < 0$ when $x > 1$
   $h''(x) < 0$ when $x > 0$
   $h''(x) > 0$ when $x < 0$.

   (a) Where is h increasing? Where is h decreasing? How do you know from the given information?

   (b) Where is h concave up? Where is h concave down? How do you know from the given information?

   (c) Where does h have critical points? Inflection points? How do you know?

   (d) Keep the function you just created. Now use a second bendable stick to graph its derivative.

   (e) At what point is h' zero? Explain what the graph of h looks like there.

   (f) What does the graph of h' look like when the graph of h is increasing (decreasing)?

   (g) Keeping both of the graphed functions in place, use a third bendable stick to graph the derivative of h'. This is also h'', the second derivative of h.

   (h) What does the graph of h'' look like when the graph of h' is increasing (decreasing)? What can you say about the shape of h at the same location?

   (i) At what point is h'' zero? What can you say about the graph of h at this point? What can you say about the graph of h' at this point?

2. Create any function on your xy-coordinate system using one of the bendable sticks. Then use another bendable stick to graph a function that is different from the first one but has the same derivative. How are these two graphs related? How do you know that they have the same derivative?

2.6 Spread the Word: Modeling Logistic Growth

Penny Dunham, Bryn Mawr College

Concepts Taught: logistic growth model, curve fitting

Activity Overview

This activity uses a random number generator and a simple set of rules to simulate the spread of a rumor throughout a small population and gives students a genuine feel for the concept of logistic growth with its classic S-shaped pattern. Students can see how the rumor starts moving slowly through the class, picks up speed as the number of people who know it reaches a critical mass, and then slows down as the number of uninitiated declines. The simulation also provides context to discuss features such as the inflection point and carrying capacity.

Supplies Needed	Class Time Required	Group Size
Random number generator* Calculator or CAS for curve fitting	15 minutes for activity 15 minutes for discussion	Whole class (15-30 students)

*For a class of size n you can use the command randInt(1,n,k) on a TI-84 for a list of k random integers between 1 and n. Alternatively, one could draw numbered slips of paper (with replacement) from a bag.

Running the Activity

To introduce the activity, provide each student with a handout, and tell the class that one person in the room is privy to an exciting rumor and their task is to track the progress of the rumor over the next few "days." Explain that we need to apply a few conditions for how the rumor spreads. (These conditions use the language of a disease model.)

1. The rumor is 100% "contagious"; anyone who hears it is in the know and can pass the secret to others.

2. Students are immune to repeat "infection"; anyone with the information can't learn it anew.

3. Contact is limited; each person who knows the rumor can tell only one person each day.

Start the activity by asking the students to stand up and count off so that each person in the room has a number. List the numbers on the board; then use a random number generator to select the student who first heard the rumor. Because each student can only tell one person the rumor each day, continue to generate enough random integers to match the number of students already in the know from the previous round; each new number indicates a student exposed to the news that "day." To create a physical representation of the rumor spreading through the class, each person sits down upon first hearing the news (i.e., when her or his number first appears). Thus, we can see at a glance how many haven't yet heard the news. The simulation ends when all students are seated.

During the activity, the first student selected tracks the movement of the secret by marking off new numbers on the board and making a chart for the total number who have heard the secret at the end of each "day". It takes about 5 minutes to explain and set up the activity, 10 minutes to run the simulation in a class of 25 students, and 15 minutes to analyze the results.

Table 2.1: Sample record of secret's spread.

Day	New random numbers	Total in the know	Daily change
1	6	1	
2	18	2	1
3	7, 8	4	2
4	4, 13, 23, 24	8	4
5	9, 11, 19, 20, 25	13	5
6	1, 2, 3, 10, 14, 22	19	6
7	5, 16, 17	22	3
8	15, 21	24	2
9	none	24	0
10	12	25	1

Table 2.1 shows the results of a simulation for a class of 25. In this case, Student #6 first heard the information on Day 1 and told #18 on Day 2. On the third day, those two students told #8 and #7, respectively, for a total of four students in the know. Condition 2 means that a number is "out" the first time it appears, so repeated numbers don't increase the total in the know. We can see the effect of this rule on Day 5. Although eight numbers were generated for those who knew the rumor on Day 4, only five were new. Thus, the count increased from 8 to 13 that day. Similarly, there was no increase in the total on Day 9 because no new numbers appeared.

After the simulation, students make a scatter plot for the data with the daily number of students in the know on the y-axis and the number of days since the rumor began on the x-axis (Figure 2.7). The class can readily see the changing rates reflected in the daily differences and the S-shape of the graph and relate those features to the physical experience of watching the rumor spread during the simulation. Following a discussion of changes visible in the graph and the roles of the inflection point and carrying capacity, students may use software to generate a logistic regression equation for the data.

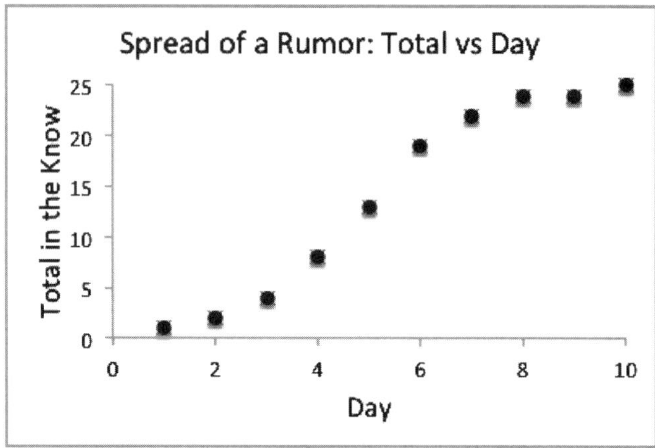

Figure 2.7: Scatter plot of the spread of the rumor.

Suggestions and Pitfalls

This activity usually runs very smoothly. Students easily relate to the scenario, and the conditions seem reasonable. I have also used a disease scenario, but it is harder for students to accept the third condition.

The physical aspect of the simulation is an important part of the activity; it offers a welcome change from sitting as well as a visual tracking as the rate of spread quickens then slows. Students remember how full the room looked in the early rounds, with nearly everyone standing, and how isolated the last few holdouts appeared in the final "days."

Another important factor is class size. With 15 or fewer students, the process terminates too quickly and often fails to produce the characteristic S-curve. In a small class, each student could take two numbers and raise both hands, lowering one when its number is called. For a large class, I recommend selecting 25 students to run the simulation to prevent having to generate long lists of random integers.

If using this activity in integral calculus, students can apply integration methods for solving logistic growth problems.

I sometimes have extended the activity to a writing project on the effects of altering the initial conditions. By changing the number of people who knew the rumor on Day 1, the number of others that one could tell on a given day, or the class size, groups made conjectures and then repeated the exploration with their own simulations. Using lists of numbers, they generated random integers to track the spread of the rumor through each "class" under different conditions and then presented their findings in a report.

Spread the Word: Modeling Logistic Growth – Class Handout

Suppose one person in our classroom has heard an exciting rumor. Starting with the one person who knows the secret on Day 1, we can track the progress of the rumor as it spreads throughout the room over the next few "days." We shall limit the model with some conditions on the way the rumor spreads.

The rumor is 100% "contagious"; anyone who hears it is in the know and can pass the secret to others.

Students are immune to repeat "infection"; anyone with the information can't learn it anew.

Contact is limited; each person who knows the rumor can tell only one person each day.

1. First, we will assign a number to each class member. Record your number here: ________.

2. Let n represent the number of students in the room and k be the total number of students who have already heard the rumor at the beginning of each day. For our class, $n =$ ________. Briefly describe how we will generate k random integers to see who has heard the rumor at each stage.

3. Use the chart below to track the progress of the rumor as we generate numbers for each day and watch the rumor spread. Continue until everyone has heard the news.

Day	New random numbers	Number in the know	Daily change
1		1	
2			
3			

4. Make a scatter plot of the data with the daily number of students in the know on the y-axis and the number of days since the rumor began on the x-axis. Describe the shape of the graph and relate it to the rate at which the rumor spread during the activity. How does the shape reflect the daily differences from the chart?

5. Why did the graph level off as the days increased? (Hint: Look up "carrying capacity.")

6. Use logistic regression to plot a curve to fit the data. How well does your model fit the observations?

7. Locate an inflection point on the graph. How does it relate to the pattern you observed in class and the numbers in the chart?

8. Summarize your observations about logistic growth.

2.7 Maximizing the Area of a Fenced in Region Using Bendable Sticks for Constraints

Julie Barnes, Western Carolina University

Concepts Taught: optimization, constraints

Activity Overview

After students are comfortable finding maxima and minima of single variable functions, they are often asked to use this information in optimization word problems. As with many word problems, students may have difficulty translating the situation described into an equation so that they can apply calculus techniques.

In this activity, students are presented with a standard fence problem where they have a fixed amount of fencing and need to maximize a rectangular area that will be built adjacent to a wall, therefore eliminating the need to put a fence on the side where the wall is located. Instead of trying to write equations immediately, students use bendable sticks to create many possible fence configurations; measure the length, width, and area of each configuration; and then compare the data generated to answers they compute with calculus. This provides students with a more concrete way to see the problem. It also attaches a physical meaning to the term *constraint*, which in this case is the length of the bendable stick.

Supplies Needed (per group)	Class Time Required	Group Size
1 bendable stick* 1 ruler	15 minutes	2-3 students

*Two brands of easy-to-use bendable sticks that work well are Wikki Stix® and Bendaroos® which can be found online and in some craft stores. Pipe cleaners can also be used.

Running the Activity

Provide groups with a handout, a ruler, and one bendable stick. Tell students that we are doing a small model of a real problem. The bendable stick represents a fence that needs to enclose an area next to a building. Because the building is on one side, there only needs to be a fence on three sides of the enclosed region. Have students create possible fences for the problem by bending the stick appropriately. Several examples of fence configurations are shown in Figure 2.8. For each possible fencing configuration, have students measure the length and width, calculate the area, and record their data on the handout.

Figure 2.8: Several possible fences created from bendable sticks.

Once everyone has at least three or four different regions measured, compile a class data table on the chalkboard. Have students scan the data to identify the maximum area and check to see if it meets the constraints.

Have students follow the rest of the handout as it guides them through the standard calculus computations and asks them questions about their observations.

When everyone has finished their calculations, lead a class discussion about any observations made. Ask questions such as, "Did our computations produce the dimensions we generated by hand? If not, why not?" Or, "Did we generate anything by hand with an area larger than the maximum computed by calculus?"

Suggestions and Pitfalls

Students are often surprised that the same constraint, i.e., length of a bendable stick, produces a wide range of areas. Encourage students to create vastly different regions with their sticks so they can see the diversity of areas computed.

Typically, there will be some data in the class list that produce an area that is larger than the maximum area allowed by the constraint. This usually comes from rounding while measuring. If this occurs, it is best to give students a chance to recognize that the constraint was broken by rounding. You could tell them that the fence will be missing a few links and a dog could escape, or something of that nature; therefore, it is important to meet all constraints.

Maximizing the Area of a Fenced in Region – Class Handout

Suppose that you need to enclose a rectangular area with a fence along the side of a building. Since the area you create will be along the side of a building, you only need a fence on each of three sides. Your job is to model this situation with a bendable stick.

1. Assume that the bendable stick provided to you is a small scale model of your fencing material. Sketch a diagram of the region in question and label the sides as length and width, with length being the side opposite the building. Use your bendable stick to generate a variety of possible fence configurations. For each fencing possibility, measure the length and width and enter the data in the table below.

Bendable stick is _______ cm long.

Length	Width	Area

2. Look at your table and estimate what you believe the maximum area is.
3. Your class will be collecting data from everyone's fences. Record your data along with your classmates, and look at the collection of data.
   (a) Which class measurements seem to create the maximum area?
   (b) Are there any constraints to the problem? That is, are there any limitations on how much fencing you can use or what shape the enclosed region must be?
   (c) Are there any data points on the class list that do not coincide with the constraints of the problem? How can you tell?
4. If there is a constraint, write an equation that describes the constraint.
5. Write an equation for the area in terms of one variable.
6. Use calculus to determine the dimensions of the fenced in region of maximal area.
7. How does your answer from calculus compare to the data generated at the beginning of class?

2.8 The Optimal Origami Box

Shelly Smith, Grand Valley State University

Concepts Taught: modeling, optimization, constraints

Activity Overview

Students often find it challenging to visualize three dimensional objects when solving optimization applications. In this activity, students fold origami boxes of various sizes, gather data, and predict the height of the box with the maximum volume. Students then use the location of the creases of the unfolded boxes to create a theoretical model for volume as a function of height, and calculate the maximum of the function. The theoretical model can be extended to boxes folded from a square sheet of paper of any size. This activity, adapted from [2, pp. 98–103], is appropriate for lessons on modeling and optimization in college algebra or calculus.

Supplies Needed	Class Time Required	Group Size
Folding instructions Square scrapbooking paper*, 2 sheets per student 3 sample boxes** Rulers, one per student Graphing technology (optional) Projector (optional)	50 minutes in class plus take-home assignment	3-4 students

*Scrapbooking paper is $12'' \times 12''$, and the large size makes it easy to fold boxes with different heights.
**Before running this activity, make sample boxes to show the students. First, use the directions in Figure 2.9 to fold a standard masu box. Then, by modifying Steps 3 and 6, fold two additional boxes of varying heights.

Running the Activity

At the beginning of this two-part activity, each student should have a copy of the handout, two sheets of square paper, and a ruler. In the first part of the activity, students create boxes of varied dimensions. Students will then use their data to make predictions about the relationship between the height and volume of a folded box. In the second part of the activity, students will use analytical methods to determine the dimensions of a folded box of maximal volume.

To get started, show the students your premade standard masu box. Then, provide students with the folding instructions for the Japanese masu (Figure 2.9); to keep everyone together, it can help to project the instructions onto a screen. Help the students follow the instructions together to fold the first box. After the students make a standard Japanese masu box, have them measure the dimensions of the box and calculate the volume, recording their answers on their handouts.

Now show students your additional premade boxes. Ask the students to conjecture how changing the depth of the folds made in Steps 3 and 6 of the folding directions (Figure 2.9) will change the height of the box. Then ask the students to predict the effect that increasing or decreasing the height of the box will have on the length and width of the box, and conjecture if this will result in increasing or decreasing the volume of the box. For the remainder of the activity, have the students form into groups, and then have each person in a group fold a box with a different height. As the group gathers data from all of their boxes, they should create a table recording the dimensions and the resulting volume of each box. In groups or as a whole class, plot the data for volume as a function of height, draw a curve to fit the data, and predict the optimum height and maximum volume of a box. If appropriate for your class, have students use technology for plotting.

The second part of the activity centers around the development of a theoretical model for the box dimensions; the questions on the handout serve as a guide for the class discussion as well as individual discovery. As a class, review the formula for the volume of a box with a square base, and discuss the need to find a relationship between the length and height of the box so that you can write volume as a function of just the height. Have the students each unfold one box and identify the creases that outline the base and sides of the box. As a class, discuss how the length of the dashed diagonal lines on the crease diagram (shown in the handout) are related to the size of the paper $\left(D = 12\sqrt{2}''\right)$. Identify the segments of the diagonal lines that represent the height and length of the boxes and the relationship between the height and length of the box and the diagonal of the paper $(D = 2L + 4H)$. Give the students the opportunity to determine that they can solve this equation for L, and use substitution to write the volume of the box as a function of its height.

Completing Problems 1-5 is a reasonable goal for a 50-minute class, and the remaining problems can be assigned for completion outside of class. By the end of class, the students should be ready to calculate the optimal height $\left(H_{opt} = \sqrt{2}''\right)$ and maximum possible volume $\left(V_{max} = 32\sqrt{2}''\right)$ of a box using graphing or differentiation. The last two problems on the handout are worded to allow flexibility for the level of the course. If you would like them to solve the optimization problem as homework in a particular way (graphically, computationally, or analytically), be sure to guide the discussion accordingly before the end of class.

Suggestions and Pitfalls

Students often need guidance when folding the first origami box and when determining how to modify the folding instructions to change the height of the box. The height of the standard box is $2.125''$. If you want the groups to fold a specific set of heights for Problem 2, assign these heights. In Problem 3, students could also use the data and technology to find a cubic model predicting the volume of a box. In Problem 5, unfolding boxes with different heights helps demonstrate that the relationships in the crease patterns are the same for all boxes. In Problem 6, you may want to include the equation $D = \_\_\_L + \_\_\_H$ as a hint. Note that the relationship $D = 2L + 4H$ is independent of the size of the paper, so this activity can easily be extended to include a class discussion of how to modify the formulas for other size paper, resulting in an optimal height of $D/12$.

References

1. Ten O'Clock Toast Time, Origami Gift Boxes, tenoclocktoasttime.wordpress.com. Accessed August 2014.

2. A. Tubis and C. Mills, *Unfolding Mathematics with Origami Boxes.* Key Curriculum Press, 2006.

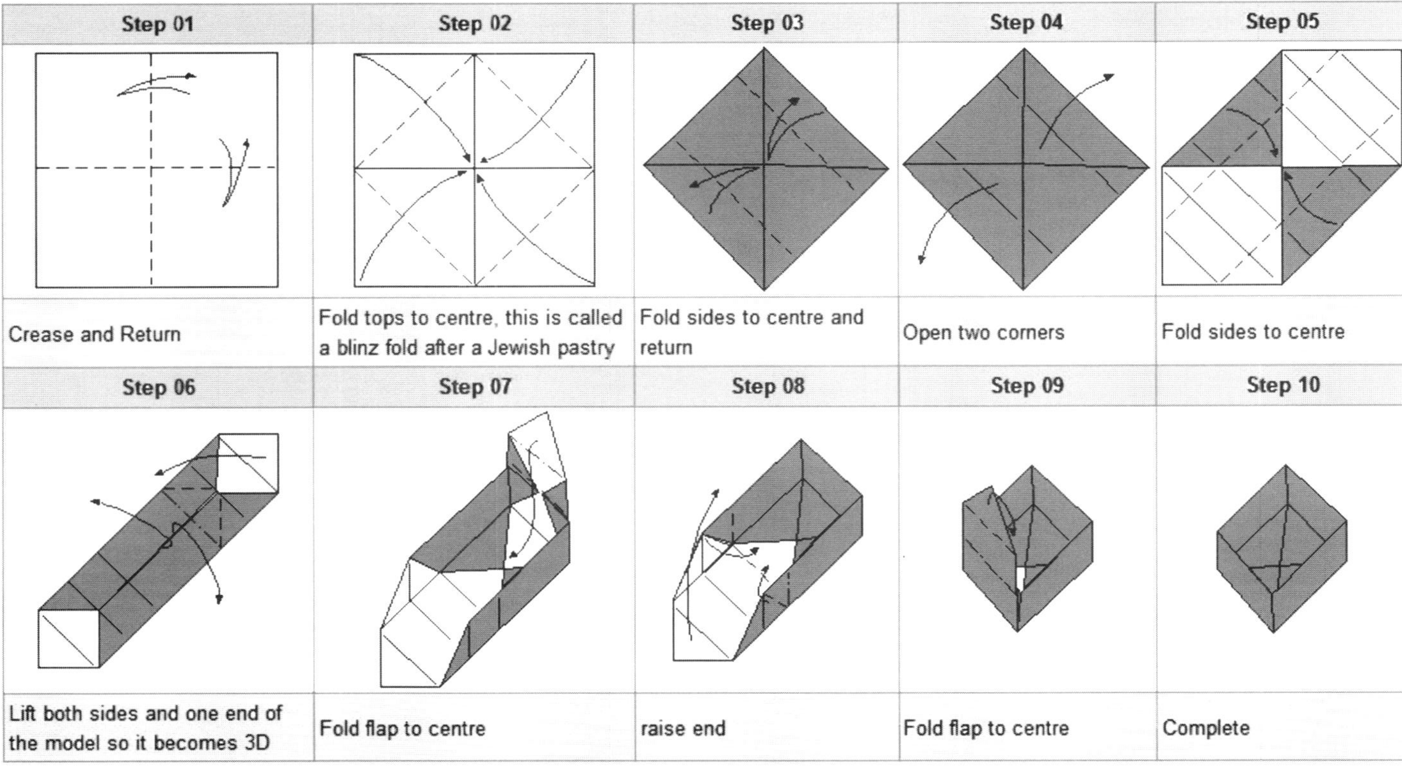

Figure 2.9: Japanese masu box folding instructions from the website tenoclocktoasttime.wordpress.com [1].

The Optimal Origami Box – Class Handout

Gathering data and making predictions

1. Using your $12'' \times 12''$ square sheet of paper, fold a box following the instructions given. Note that since our paper is square and our folding process is symmetric, the base of the box is also square. Measure the dimensions of your box, and calculate its volume. Include units on your answers.

2. In your group, fold boxes of different heights by adjusting the height of the sides you fold in Steps 3 and 6. Measure and calculate the volume of each box, and create a table of your data.

3. Create a scatterplot of your data and draw a smooth curve to connect the points. Predict the height of the box that will have the maximum possible volume. Based on your prediction, what is the maximum possible volume for your origami box?

Creating a theoretical model

4. What is the formula for the volume of a box with a square base?

5. Unfold one of the boxes that you made so that you can see the creases. They should look similar to the diagram below. Trace the creases that outline the base of your box and the creases that outline the sides. What is the length of the dashed diagonal lines that have been added?

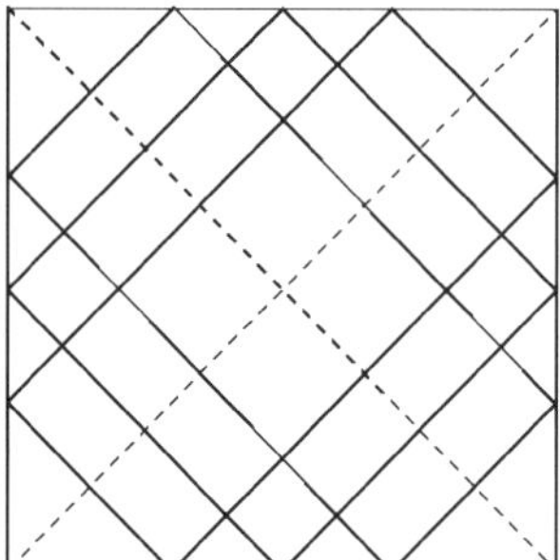

6. Identify segments of the diagonal lines on the crease diagram that can be used to measure the length and height of the box. Use these segments to find an equation that relates the length of the diagonal, D, to the height, H, and length, L, of the box. How can we use this equation to eliminate L from the volume formula, so that the volume is a function of the height of the box?

7. What mathematical approach do you suggest for determining the optimal height and maximum volume of an origami box?

8. Use the volume function you created to determine the maximum possible volume for your origami box. How does this result compare to your prediction in Problem 3?

Chapter 3

Integral Calculus

3.1 Chewing Gum Riemann Sums

Julie Barnes, Western Carolina University
Jessica Libertini, Virginia Military Institute

Concepts Taught: Riemann sums, area approximation

Activity Overview

Although area approximations using Riemann sums – including left-endpoint, right-endpoint, and midpoint methods – may seem self-explanatory, their early placement in the learning of integral calculus provide an opportunity to slow down and make sure students have a chance to internalize these concepts and their foundational importance to the study of integration. In this activity, students align sticks of gum along a curve creating a visual representation of a set of Riemann sums, exploring the common endpoint methods as well as different limits of integration. This activity also gives instructors a means to quickly confirm and correct student understanding.

Supplies Needed (per group)	Class Time Required	Group Size
5-stick pack of chewing gum*	15-20 minutes	1-3 students

*The chewing gum should be the classic 5-pack stick gum.

Running the Activity

Give each group a pack of gum and a copy of the handout which includes a graph of an increasing function. (The online handout is in gum-width sized units.) Have groups begin working through the handout. Problem 1 asks students to set up their gum to show the area represented by a left-endpoint Riemann sum over the five unit interval from 0 to 5. Once they have their gum aligned, have them use a piece of paper to cover the portion of the gum that hangs below the x-axis, as the overhang does not contribute to the area and its presence may confuse weaker students. As you walk around the room, verify that they have properly aligned the gum as shown in Figure 3.1; you can also ask them to check with a neighboring group to compare. Problem 1 goes on to ask students to determine the relevant area approximation and decide whether it is an overestimate or underestimate.

Have the students continue to work through the rest of the handout as you walk around the room verifying their gum placement on each problem and providing hints or suggestions as needed. In all cases, make sure it is clear to the students that the portion of the gum falling below the x-axis is not part of the area in question.

Problems 2 and 3 on the handout ask students to repeat the approximation for this region using right-endpoints and midpoints for the Riemann sums respectively, while Problem 4

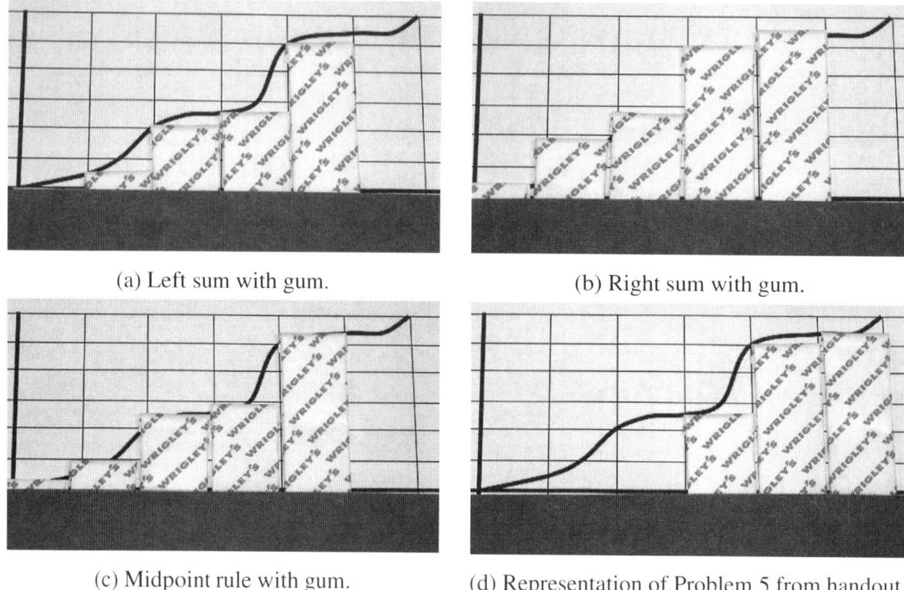

(a) Left sum with gum.

(b) Right sum with gum.

(c) Midpoint rule with gum.

(d) Representation of Problem 5 from handout.

Figure 3.1: Some examples of solutions to problems from the handout.

has them repeat the process using any point in each subregion. In Problem 5, as the students approximate the area between $f(x)$ and the x-axis from $3 \leq x \leq 6$, it is important to observe their work to make sure that they understand how to address this change in limits. Finally, Problems 6 and 7 ask students to extend the concepts covered in this activity to more generalized functions.

Suggestions and Pitfalls

One of the greatest benefits of this activity is that you, as the instructor, can easily assess which students understand the ideas of Riemann sums and which students need more personalized attention. Due to the nature of the activity, stronger students are able to continue working on the activity while you spend more time assisting students who are struggling.

Often students have a hard time understanding how to start the midpoint method, but it helps if you ask them to explain verbally how they did left- and right-endpoint methods. A typical student response is, "We moved the gum up or down until the left (or right) edge of the gum touched the function." Then, with a little guidance, students often make the leap that for the midpoint method, they will need to slide the gum up or down until the middle of the gum touches the function.

Due to the scale of this activity, it helps to keep the groups very small so that each person is actively involved in moving the gum. Students who simply watch the activity often believe they understand, but those who directly manipulate the gum are more likely to internalize the lesson. While the activity may seem trivial, some students really struggle with this concept, so it is beneficial to set aside more time than you think your students will need. Spending time on this early can save a lot of time later.

Chewing Gum Riemann Sums – Class Handout

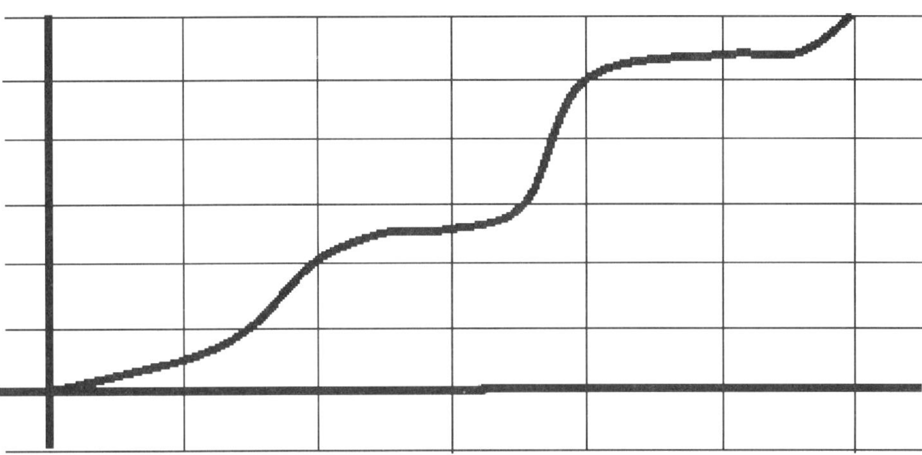

1. Using your gum, represent the area between $f(x)$ and the x-axis from $0 \leq x \leq 5$ with a left-endpoint Riemann sum. Once the placement of your gum has been verified, use a piece of paper to cover the portion of gum that lies below the x-axis, and use the visible portion of the gum to approximate the area between $f(x)$ and the x-axis from $0 \leq x \leq 5$. Is your answer an overestimate or underestimate of the actual area?

2. Using your gum, represent the area between $f(x)$ and the x-axis from $0 \leq x \leq 5$ with a right-endpoint Riemann sum. Once the placement of your gum has been verified, use a piece of paper to cover the portion of gum that lies below the x-axis, and use the visible portion of the gum to approximate the area between $f(x)$ and the x-axis from $0 \leq x \leq 5$. Is your answer an overestimate or underestimate of the actual area?

3. Using your gum, represent the area between $f(x)$ and the x-axis from $0 \leq x \leq 5$ with a midpoint Riemann sum. Once the placement of your gum has been verified, use a piece of paper to cover the portion of gum that lies below the x-axis, and use the visible portion of the gum to approximate the area between $f(x)$ and the x-axis from $0 \leq x \leq 5$.

4. Can you choose *any points* between the left and right endpoints such that the resulting estimate is bigger than your overestimate or smaller than your underestimate? Explain your answer.

5. Using your gum, approximate the area between $f(x)$ and the x-axis from $3 \leq x \leq 6$ with a left-endpoint Riemann sum. Once the placement of your gum has been verified, use a piece of paper to cover the portion of gum that lies below the x-axis, and use the visible portion of the gum to approximate the area between $f(x)$ and the x-axis from $3 \leq x \leq 6$. What do you need to do differently from Problem 1 to solve this problem?

6. Consider a new function, $g(x)$, which is a decreasing function. Which will give an overestimate, a left-endpoint Riemann sum or a right-endpoint Riemann sum? Explain your answer.

7. Consider another new function, $h(x)$, that has both increasing and decreasing regions. Can you state with certainty whether the left-endpoint Riemann sum will be an overestimate or underestimate? Why or why not?

3.2 Paper Shredder Riemann Sums or Cut the Bunny

Carolyn Yackel, Mercer University

Concepts Taught: Riemann sums, area approximation

Activity Overview

Students have a hard time connecting the areas of rectangles used for standard Riemann sums with areas of actual rectangles that one might naturally use to approximate the area between a function and the x-axis. As a consequence, students asked to use a right- or left-hand sum sometimes insert wild numerical approximations of functional height that seem unrelated to the actual height of the function on the interval. The purpose of this realistic activity is to have students derive for themselves a reasonable method for approximating the area of a region with curved sides. Students are asked to approximate the area of a paper shape; to do this, they run the shape through a paper shredder and add up the approximate area of each strip. Via the activity the students connect the various standard methods of approximating the area of a region with each other and with the formula as given through notation. In addition, the activity intuitively clarifies the role of the limit of the Riemann sum in passing to the exact value of the area.

In addition to providing a context in which students may originate for themselves the notion of Riemann sums, your role as instructor is to support student investigation through asking appropriate guiding, deepening, and rationale questions. Finally, you can use this activity to create a classroom event worth referencing again and again by weaving together the results and observations of students to create the basis for the theory of Riemann sums.

Supplies Needed	Class Time Required	Group Size
Half sheet of colored paper per group 1 pair of scissors per group 1 strip-cut paper shredder*	20 to 30 minutes depending on discussion length	3-4 students

* Be sure to use a strip-cut paper shredder as opposed to a cross-cut paper shredder. Strip-cut paper shredders may be difficult to find in stores but can be ordered online.

Running the Activity

Put students into groups of three or four and provide each group with a handout, a half sheet of colored paper, and scissors. Have each group cut a bunny or other abstract shape out of their paper. For best results, require shapes with some curved edges, and show them a sample shape indicating size and basic form. Have students try to estimate the area of the shape using ad hoc methods. Observe this process and note the methods they try.

After only a couple of minutes, introduce the presence of the paper shredder as a potential tool. Cut the bunny: run the sample shape through the paper shredder as an example idea for how this tool might be used as seen in Figure 3.2.

Let student groups continue to work on estimating the area of their shapes. Continue to observe their processes. Eventually they will all use the shredder. Allow groups to decide how to measure and add their rectangles. Encourage groups to articulate their approaches. It is not important that groups obtain final area totals; needed results are ideas about their processes. Therefore, make sure groups have enough time to identify their process including the details of how they will approximate the area, how they will record their results, and which member is in charge of each part of the process (measuring, recording, multiplying, adding).

Finally, bring the class back together for discussion. Discuss the approach of each group, focusing not on the values of the areas but on the methods, particularly on the issues that arose once the paper shredder was used and what decisions each group made.

Suggestions and Pitfalls

Allowing only one half-sheet of paper, one pair of scissors, and one ruler (on the handout) per group forces students to work together. Additional verbal encouragement and ideas for productive collaboration may be needed.

Using different colored half-sheets of paper is useful, because often one group leaves a few strips in the bottom of the shredder by accident. The next group finds these stragglers only because they are a different color.

The purpose of asking students to use ad hoc methods is to motivate Riemann sums by causing frustration with existing methods, which is why not much class time is budgeted for that component. If you want students to be able to determine the area of their shapes in this activity using ad hoc methods, such as by dissecting it into rectangles and triangles, then Riemann sums will not be motivated.

Allowing groups to generate their own methods for determining the areas of the paper strips fosters a rich wrap-up discussion of various Riemann sums methods because the valid methods of different groups can be discussed and compared.

Brief consultations with groups during work time about their method choices help students to make coherent explanations to the whole class during the wrap-up discussion. The consultations also inform the instructor which techniques students will not have generated via the activity.

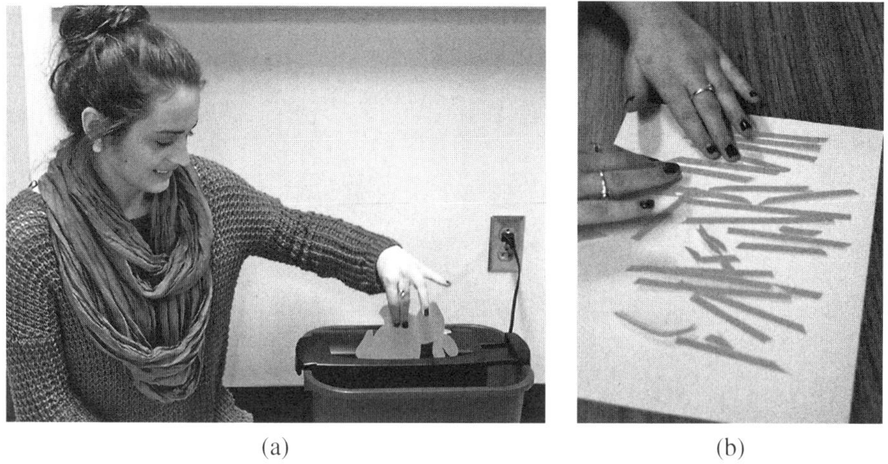

(a) (b)

Figure 3.2: (a) Shredding a paper bunny. (b) Contemplating the pieces that need to be measured.

Cut the Bunny – Class Handout

During this activity, you will work in a group to think of clever ways to approximate the area of an abstract shape that you cut out of the colored paper given to you by your instructor.

1. Use your scissors to cut out a shape with curved edges. Use the majority of the area of the paper for your shape.

2. Use the centimeter ruler copied onto the bottom of this page as you attempt to estimate the area of your shape. **Make records of all of your efforts that indicate both your ideas for finding the area and your estimation of the area.** Be sure to record any difficulties. If you get stuck, consider the following questions, not necessarily one at a time.

   - What techniques do you know for finding areas of geometric objects?
   - For what types of geometric objects do you know formulas for the area?
   - If your shape was simply comprised of these geometric objects, would you be able to find its exact area? Articulate the obstacle to finding the exact area of your shape here.
   - Are there perhaps some geometric objects that approximate the area less precisely that avoid this obstacle?
   - How can conceiving of your area as divided into regions help you in the process of determining the total area?
   - Are there special considerations or efficiencies that come from dividing the region into areas in the way that you have? (For example, in what order do you need to add your areas?)

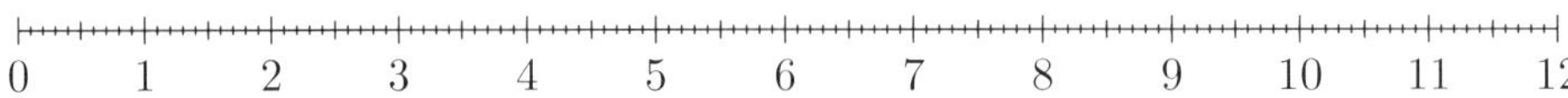

3.3 Estimating Calories in a Cookie with Riemann Sums

Julie Barnes, Western Carolina University

Concepts Taught: Riemann sums, area estimation

Activity Overview

In this activity, various estimation techniques of integration (left sum, right sum, trapezoid rule, and midpoint rule) are compared and used to generate an estimate for the surface area of the bottom of an oddly shaped cookie. At the end of the activity, students use this calculation for surface area to estimate the number of calories in the cookie. This provides students with a concrete example of how estimation techniques can be used.

Supplies Needed (per group)	Class Time Required	Group Size
1 wrapped homemade cookie of unusual shape* 1/4 inch grid paper Calculators	20-30 minutes	2-3 students

*Bake enough cookies prior to class. One thing that works really well is to purchase sugar cookie dough that breaks off into cubes, use two cubes for each cookie, flatten the dough so that it is approximately $1/2''$ thick, and arrange the dough on the pan so that the cookies will not be perfectly round when baked. After baking and while the cookies are still warm, cut them in half so that the flat edge could be lined up against an x-axis and the remaining edge is actually a function as seen in Figure 3.3. The straight edge helps students connect the activity to more general functions, especially if the students are not used to modeling real life objects. Also note that to make any calorie estimations plausible, flattening the dough to be roughly $1/2''$ before baking helps make the thickness fairly uniform. Also, using sugar cookie dough avoids complications with items like chocolate chips which have a different number of calories than the dough. Before class, cover each cookie in plastic wrap. This keeps the cookies free from germs as students are tracing them, it keeps grease off of their homework papers, and the wrap is pliable enough to still accurately trace the cookie.

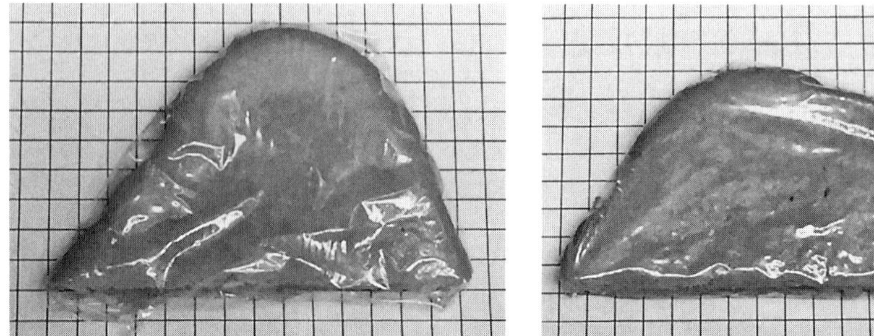

Figure 3.3: Oddly shaped cookies with one side straight, covered in plastic wrap, and positioned on $1/4$ inch graph paper.

Running the Activity

After a quick review of the appropriate integration approximation techniques, have students get into groups. Provide each group of students with a cookie, $1/4$ inch grid paper, and a handout. As they follow through the handout, students make four traces of their cookie, one per quadrant of the $1/4$ inch graph paper as shown in Figure 3.3. Each traced image corresponds to one of the four integration techniques (left sum, right sum, trapezoid rule, and midpoint rule). Then, for each diagram students sketch boxes to represent the method they are using and then use the given method to estimate the surface area of the bottom of the cookie in terms of the number of $1/4$ inch boxes used. Since this is an unusual unit of measure, the handout has students convert from $1/4$ inch boxes to square inches; students need to recognize that the area of one box is $1/16$ in^2. Once they have estimated the surface area using each of the different methods described in the handout, Problem 5 asks them to determine which of the methods is the best estimate for the surface area and consider errors made in estimating the surface area.

The last problem asks students to use their best surface area estimate to approximate the number of calories in the cookie. To do this, students will need an estimate of the number of calories a cookie has per square inch, as provided on the handout. Then they should address possible errors in estimating calories this way. For example, the cookies may not be uniformly thick.

Suggestions and Pitfalls

In order to complete the last problem on the handout, students are given an estimate of the number of calories a cookie typically has per square inch of the bottom of the cookie's surface. However, this value could vary depending on the brand of dough and how they bake in your oven. If using prepackaged sugar cookie dough, the calories per cubic inch can be computed from the given nutritional information on the package. Typically it is around 14 cal/in^2, but you may want to calculate it yourself. Since computing this information is not related to the goals of the activity, it is best to make this estimate yourself in advance and provide it to the students or just use 14 cal/in^2 as given on the handout.

For more of a challenge, use the whole cookie instead of providing a flat edge and adjust the directions at the top of the handout accordingly. Store-bought cookies could also be used as long as they are unusually shaped, and do not have areas easily computed with basic geometry.

Depending on the topics you are covering, it is easy to modify which methods you want students to use. For example, they could use the points from their diagram to apply Simpson's rule.

Estimating Calories in a Cookie with Riemann Sums – Class Handout

Make four traces of the cookie, one per quadrant of the $1/4$ inch graph paper. Each time you trace the cookie, line up the straight edge with a horizontal line and the left corner touching a vertical line. The horizontal edge will be your x-axis, and the line the cookie touches on the left is the y-axis.

1. On the first sketch of the cookie, draw in rectangles that represent a left sum. Use rectangles whose width is the width of the boxes, $1/4$ inch.
   (a) Use a left sum to calculate the number of $1/4$ inch boxes inside the curve. The units will be $1/4$ inch boxes.
   (b) Convert your answer to square inches.
2. On the second sketch of the cookie, draw in rectangles that represent a right sum. Use rectangles whose width is the width of the boxes, $1/4$ inch.
   (a) Use a right sum to calculate the number of $1/4$ inch boxes inside the curve. The units will be $1/4$ inch boxes.
   (b) Convert your answer to square inches.
3. On the third sketch of the cookie, draw in rectangles that represent the midpoint rule. Use rectangles whose width is the width of the boxes, $1/4$ inch.
   (a) Use the midpoint rule to calculate the number of $1/4$ inch boxes inside the curve. The units will be $1/4$ inch boxes.
   (b) Convert your answer to square inches.
4. On the fourth sketch of the cookie, draw in trapezoids that represent the trapezoid rule. Use trapezoids whose width is the width of the boxes, $1/4$ inch.
   (a) Use the trapezoid rule to calculate the number of $1/4$ inch boxes inside the curve. The units will be $1/4$ inch boxes.
   (b) Convert your answer to square inches.
5. Look over your four answers as well as the sketches you have drawn.
   (a) Based on your sketches, which method(s) do you believe would provide the best estimate of surface area? Why?
   (b) What possible errors do you see in using these estimation techniques?
6. Typically these cookies contain around 14 cal/in^2.
   (a) Use your best surface area estimation to approximate the number of calories in your cookie.
   (b) What possible errors do you see in estimating calories in this way?

3.4 So Many Integration Techniques... Which to Use?

Martha Allen, Georgia College & State University
Karen Bliss, Virginia Military Institute
Jessica Libertini, Virginia Military Institute

Concepts Taught: techniques of integration

Activity Overview

Students learn integration techniques one by one, and it is not uncommon for instructors to assign problems targeting one technique at a time. For example, it is easy for students to decide which technique to use when they are doing homework in the *Integration by Parts* section of the textbook. While focused homework helps students practice their new-found integration skills, it does not help them decide which technique they should use when they encounter an unfamiliar integral on an exam. This sorting activity is designed to help students develop this ability by having them focus on classifying indefinite integrals based on the integration technique needed to evaluate the integral.

Supplies Needed (per group)	Class Time Required	Group Size
Scissors	20 minutes	2 students

Running the Activity

This activity is intended to assist students in determining which among several integration techniques they should use. As such, use this activity after you have covered all of the integration techniques addressed in your course. This can be used as a part of a review session before an exam on integration techniques.

Give each group of students one copy of the handout along with a pair of scissors. Have your students cut out the tiles containing examples of integrals and the names of integration techniques. As they get started, create an empty table on the board, as in Figure 3.4.

Substitution	Integration by Parts	Trigonometric Powers and Products	Partial Fraction Decomposition	Inverse Trigonometric Function	Polynomial Long Division First

Figure 3.4: An empty table to write on the board in the classroom. Students will put the appropriate integral numbers under each integration technique.

Have students sort their tiles by first determining which integration technique they would use if they wanted to find the antiderivative and then grouping their pieces accordingly, as shown in Figure 3.5. As they work, walk around the room, providing guidance as needed. When you begin to notice that students are finishing, select students to go up to the table on the board and list integral numbers under the appropriate integration technique. For example, the $\#12$ should appear in the Substitution column because that is the technique one would use to evaluate $\displaystyle\int \frac{e^{1/x}}{x^2}\,dx$.

After students have filled in the table on the board, ask them to compare their results with those on the board. It is possible that some numbers appear in multiple columns while others have not been listed on the board at all. Facilitate a discussion among the students, allowing them to debate which integrals belong under which category and why. Continue this process, updating the board based on their discussion, until the class reaches consensus and each integral has been properly categorized.

Suggestions and Pitfalls

The handout can be modified based on the integration techniques your class studies. To keep the focus on choosing an appropriate integration technique, we suggest keeping most of the integrals on the simpler side.

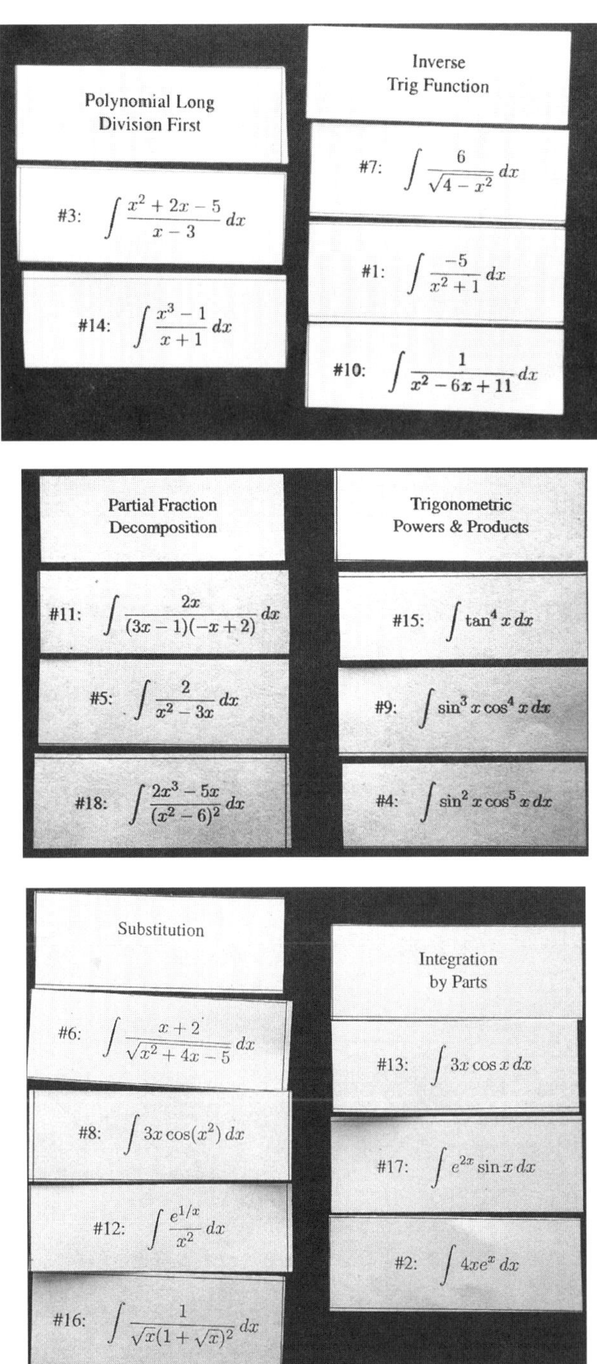

Figure 3.5: Example of completed activity.

So Many Integration Techniques ... Which to Use? – Class Handout

Cut out each of the following integrals and integral techniques. Spread the integral techniques out across your desk.

For each of the integrals, determine which integration technique you would use to find the antiderivative and place the integral next to that technique.

You do not need to evaluate any of the integrals; simply determine which integration technique applies.

#1: $\int \frac{-5}{x^2+1}\,dx$	#2: $\int 4xe^x\,dx$	#3: $\int \frac{x^2+2x-5}{x-3}\,dx$
#4: $\int \sin^2 x \cos^5 x\,dx$	#5: $\int \frac{2}{x^2-3x}\,dx$	#6: $\int \frac{x+2}{\sqrt{x^2+4x-5}}\,dx$
#7: $\int \frac{6}{\sqrt{4-x^2}}\,dx$	#8: $\int 3x\cos(x^2)\,dx$	#9: $\int \sin^3 x \cos^4 x\,dx$
#10: $\int \frac{1}{x^2-6x+11}\,dx$	#11: $\int \frac{2x}{(3x-1)(-x+2)}\,dx$	#12: $\int \frac{e^{1/x}}{x^2}\,dx$
#13: $\int 3x\cos x\,dx$	#14: $\int \frac{x^3-1}{x+1}\,dx$	#15: $\int \tan^4 x\,dx$
#16: $\int \frac{1}{\sqrt{x}(1+\sqrt{x})^2}\,dx$	#17: $\int e^{2x}\sin x\,dx$	#18: $\int \frac{2x^3-5x}{(x^2-6)^2}\,dx$
Substitution	Integration by Parts	Trigonometric Powers & Products
Partial Fraction Decomposition	Inverse Trig Function	Polynomial Long Division First

3.5 Centers of Mass of Candy Point Masses on a Plane

Erin Bancroft, Grove City College

Concepts Taught: center of mass

Activity Overview

A common application of integration is finding the center of mass of a lamina. In this activity, students physically explore and discover the formulas for finding the center of mass of point masses in the plane by building and balancing a lamina. This hands-on experience is designed to aid in student understanding of the meaning of center of mass.

Supplies Needed	Class Time Required	Group Size
1 bag each of Hershey's Kisses® and Starburst®* 1 cardboard square Cartesian plane, per group** A few rolls of clear tape	40-50 minutes	2-4 students

*Two distinct classes of non-edible objects or alternative candies can be used in place of the Hershey's Kisses and Starburst as long as they are small. You will need to provide the mass of each class of objects.

**To create the cardboard Cartesian plane, use scissors or a box cutter to cut squares from a sturdy piece of cardboard. For best results, the squares should have side length between $5''$ and $8''$. Boxes used for packaging appliances (with the plastic coating) make particularly nice squares. Cut a template out of paper, and use it to trace out the squares on the cardboard. Be careful not to include any bends or creases on the interior of each square as you want a flat, uniformly thick surface. Once you have cut out the squares, print off two versions of the Cartesian plane, one for each side of the square. A scaled-down example template is given in Figure 3.6. (A full-size version is available in the online handouts.) The "top" plane should be a standard Cartesian plane, as in Figure 3.6a, and the "bottom" plane should be flipped across the y-axis, as in Figure 3.6b. The planes should be the same size as your square with a scale running from -4 to 4 in both directions. It is helpful to have the xy-axes labeled as well as the "top" and "bottom" plane. Attach the Cartesian plane to the cardboard squares using a glue stick or spray-on glue. Make sure that the y-axis points in the same direction on both sides. An example of the cardboard plane with candy on it is given in Figure 3.7a.

Running the Activity

Pass out four Hershey's Kisses, four Starburst candies, a roll of clear tape (or several pieces), a cardboard Cartesian plane, and handouts to each group. Students should then work through the handout in their groups. They will begin in one dimension by finding the center of mass of two and three Hershey's Kisses on the x-axis and then generalizing to a formula for the center of mass of any number of point masses on a single axis. From there, they will work in two dimensions and develop a formula for finding the center of mass of any number of Hershey's Kisses placed on a Cartesian plane. After being introduced to the definition of *moment* in Problem 5, the students will go back to a single axis to determine the center of mass for points of different masses such as the Hershey's kiss and Starburst. Finally, students are asked to synthesize the results of the previous problems as they determine the center of mass of several pieces of candy placed at specific points in the Cartesian plane. The students can verify their results as shown in Figure 3.7b.

Suggestions and Pitfalls

Make sure students know that they will need all the candy for the duration of the activity. Emphasize to the students that they should think about and calculate the locations of the centers of mass by hand before trying to get the system to balance on their finger. Depending on the level of the students in the class and how much time you have, you may want to pause the group work periodically and discuss the answers to the problems on the handout as a class. This will help to prevent groups from going ahead with incorrect solutions or ideas.

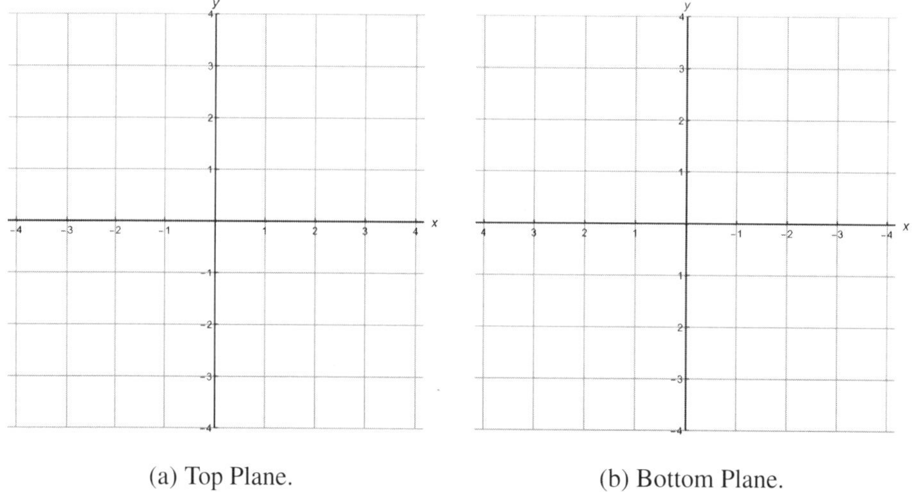

(a) Top Plane.

(b) Bottom Plane.

Figure 3.6: Cartesian plane templates.

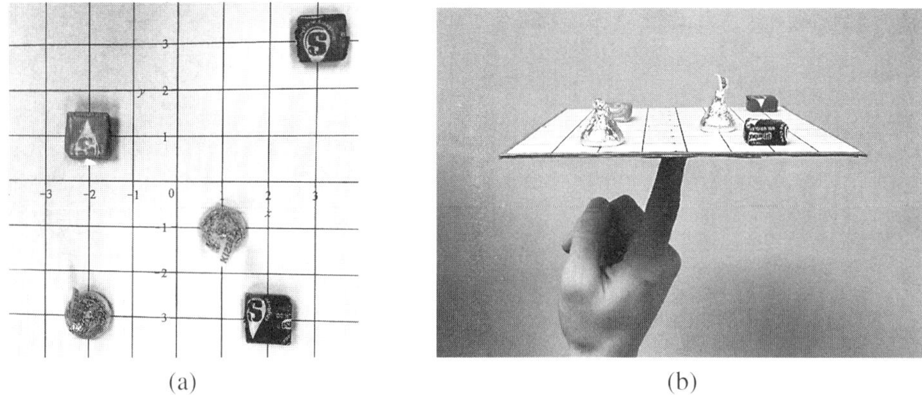

(a) (b)

Figure 3.7: (a) Placing the candy on the cardboard coordinate system. (b) Balancing the system with your finger.

Centers of Mass of Candy Point Masses on a Plane – Class Handout

1. Using tape, attach two Hershey's Kisses® on the x-axis, one at $x = -1$ and one at $x = 3$. What is the balancing point of this system? In other words, at which point on the board could you put your finger underneath and have the board balance? (We call this point the *center of mass* of the system.) Discuss among your group, and then test out your answer.

2. Continuing Problem 1, attach an additional Hershey's kiss to the point $x = -2$, and find the new center of mass of the system. Use your results to create a formula to find the center of mass of any number of Hershey's Kisses on a single axes.

3. Now attach four Hershey's Kisses to the plane, one at each of the points $(0, 0)$, $(0, 3)$, $(-3, 0)$, and $(-3, 3)$. What is the center of mass of this system? Discuss and test your result. Then, develop a formula for finding the center of mass of any number of Hershey's Kisses placed on a Cartesian plane.

4. Would your formula still apply if you used all Starburst® instead of Hershey's Kisses? What if you used a mixture of Starburst and Hershey's Kisses? Discuss.

5. The *moment* of a point mass in the plane about an axis is its mass multiplied by the distance from the axis. For a mass of 2 grams at the point $(3, -2)$ the moment of the mass about the x-axis is $-2 \cdot 2 = -4$ and the moment of the mass about the y-axis is $3 \cdot 2 = 6$. Suppose you have a Hershey's kiss at $x = -2$ and a Starburst at $x = 3$. A Starburst has a mass of approximately 5.05 grams and a Hershey's kiss has a mass of 4.65 grams. Find the center of mass of the system. Create a formula for finding the center of mass of points of different masses on a single axis. Express your answer in terms of moments if you have not done so already.

6. Take three pieces of Starburst candy. Attach them at the points $(-2, 1)$, $(2, -3)$, and $(3, 3)$. Take two Hershey's Kisses; attach them at the points $(-2, -3)$ and $(1, -1)$. Think about how you can use your answers from the previous problems to develop a formula and then use it to find the center of mass of this system. Test out your answer by seeing if your system will balance when you place your finger underneath the point you found.

3.6 Volume Estimations with Fruit Cross Sections

Elin Farnell, Colorado State University
Marie A. Snipes, Kenyon College

Concepts Taught: Riemann sums, solids of revolution, volume estimations, modeling

Activity Overview

Students sometimes struggle with visualizing three dimensional solids, including solids of revolution. In this two-day activity, students create a print of a roughly radially symmetric fruit and use Riemann sums to estimate its volume. In addition to giving students an opportunity for tactile learning related to solids of revolution, this project improves student understanding of functions of unknown formulas and stresses the use of correct notation as students construct a Riemann sum.

Supplies Needed	Class Time Required	Group Size
One fruit or vegetable per group, sliced in half* Tempera paint and paintbrushes Paper and rulers	Day 1: 15-20 minutes Day 2: 35 - 50 minutes	2 students

*Select fruit that is roughly radially symmetric such as lemons, oranges, apples, or mushrooms. Prior to the activity, slice each fruit along a plane containing the axis of rotation as seen in Figures 3.8a and 3.9a. Be sure to keep matching halves together.

Running the Activity

For the purposes of this discussion, we will refer to the use of lemons and will suggest potential variations later in the Suggestions and Pitfalls section. At the start of the activity, student pairs select a pre-sliced lemon, paint one or both faces of the lemon halves using tempera paint, and make prints of their lemon faces on paper until they feel that they have a good representation of the cross section. Student pairs may get more accurate prints if one person holds the lemon half while the other places the paper on top and presses down; this gives a cleaner print than pressing the lemon down onto the paper. See Figures 3.8a and 3.8b for an example of a painted lemon and its corresponding print. Prints need to dry (for about 15-20 minutes) before they can be traced, so students should make their prints at the end of the first class day. On the second class day, have each pair of students collect their prints and work through the handout. Problems 1-6 guide students through tracing their prints and using them to develop a Riemann sum for the solid of revolution. Problems 7 and 8 have students estimating the volume of their fruit and discussing possible sources of error. See Figure 3.8c for a possible setup for a volume calculation.

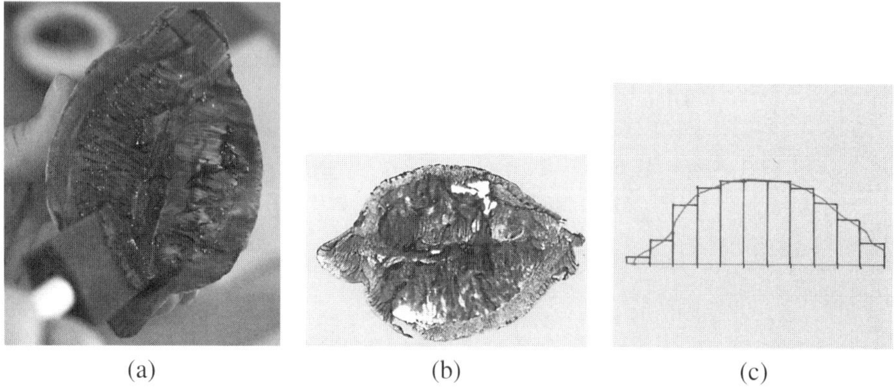

(a) (b) (c)

Figure 3.8: (a) A painted lemon. (b) A print. (c) A setup for a volume calculation using the disk method.

Suggestions and Pitfalls

While we expect most students to use the disc method to approximate the volume, it is also reasonable for students to use the shell method.

We like to emphasize that they will get a more accurate approximation by choosing an appropriate point in each interval rather than always using a left or right endpoint. This activity is an ideal example of a situation in which one might want to choose evaluation points differently within each subinterval in order to obtain a reasonable estimate.

Many students want to fit a function to the curve they trace and use the function formula to evaluate their Riemann sum. We highlight the fact that no formula is needed, and that students should instead make measurements to use as their rectangle heights at the various evaluation points.

To further enhance this activity, consider providing tracing paper to the students. Then students can trace the entire cross-section, fold it in half, and determine where asymmetries or inaccuracies might exist.

Students may be interested in also estimating the volume using water displacement; this provides a common sense check to see if the answers they obtained using Riemann sums are reasonable.

More challenging questions can be asked. For example, if students paint the rind of the lemon instead of the entire face of the cross-section, they can estimate how much rind the lemon has. If a variety of fruits are used, the class could make comparisons of the ratio of rind to fruit. See an example print for a lemon rind and a minneola rind in Figures 3.9a, 3.9b, and 3.9c. Similarly, avocados could be used, and students could calculate the volume of the avocado with the pit removed. In a slightly different vein, if one uses an apple or a mushroom as the object of study, there are interesting computations that arise as a result of the apple shape and the mushroom gills. See Figures 3.9d and 3.9e for prints of an apple and a mushroom.

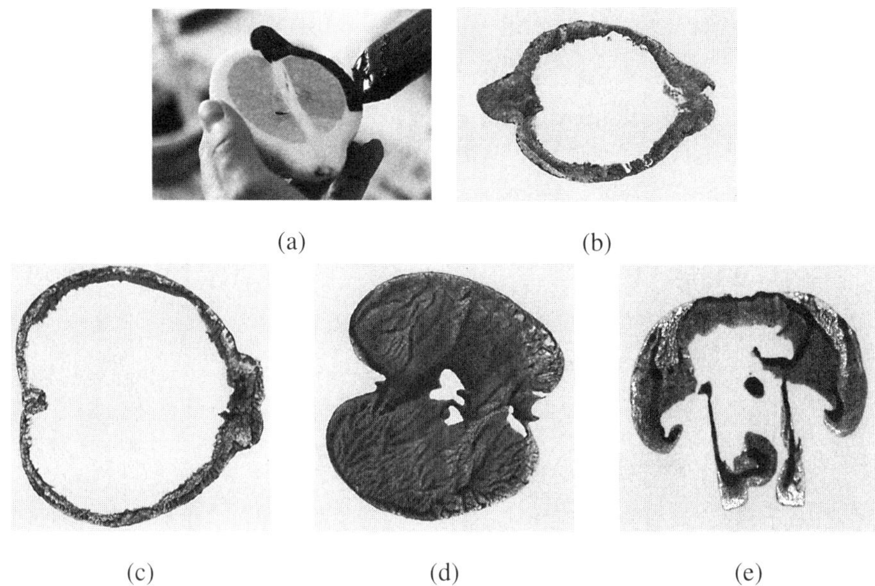

(a) (b)

(c) (d) (e)

Figure 3.9: (a) Painting a lemon rind. (b) Print of a lemon rind. (c) Print of a minneola rind. (d) Print of an apple. (d) Print of a mushroom.

The idea for this activity arose out of a paper about using pottery for a similar exercise. We modified the content to make it accessible to instructors who may not have pottery studios. Those interested in the pottery activity are referred to our article in *PRIMUS* [1].

Reference

1. E. Farnell and M. A. Snipes, Using the pottery wheel to explore topics in calculus, *PRIMUS* 25 no. 2 (2014) 170–180.

Volume Estimations with Fruit Cross Sections – Class Handout

1. Use the axes below to make a tracing of half of a print of your fruit. The axis of revolution for the fruit should coincide with the x-axis.

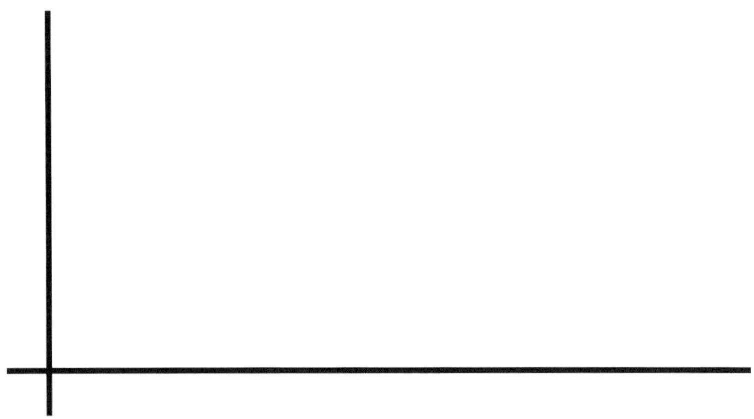

2. Choose a partition for the relevant interval and label the partition points $x_0, x_1, \ldots, x_n$ on the graph above. You may use any n greater than or equal to 10.

3. List the values of Δx_k for k from 1 to n. (This may be a single value if you used a uniform partition.)

4. Choose and label evaluation points $x_1^*, x_2^*, \ldots, x_n^*$. Draw in rectangles with height $f(x_k^*)$. Use a ruler to measure $f(x_k^*)$ in each case.

5. For a subinterval of your choice, use the space below to draw the the shape that will result from revolving the given rectangle around the x-axis. Label Δx_k and $f(x_k^*)$ on your drawing. For the chosen subinterval, calculate the volume of the revolved rectangle.

6. On a separate sheet of paper, make a table with a column for each of the following and fill out the table for k from 1 to n: $\Delta x_k, f(x_k^*)$, and the volume estimate from the kth slice.

7. Compute an estimate of the total volume of your fruit.

8. Discuss what sources of error might have contributed to inaccuracies in your volume estimate.

3.7 Fun with Infinite Series

Jennifer Hutchison, Cedarville University

Concepts Taught: geometric series, series convergence, modeling

Activity Overview

Series convergence and divergence tend to be counterintuitive concepts for many students. Adding up infinitely many numbers to obtain a finite sum, but only sometimes, can seem like sophistry, and that makes sense: the dividing line between converging and diverging is very fine. Geometric series are an excellent example of this. They are especially useful for examples since there is an easy formula to calculate the sum of a convergent geometric series. This activity includes three concrete examples of infinite processes that illustrate both convergent and divergent series: cutting paper in half, the Koch snowflake curve, and a bouncing ball. A class could do one, two, or all three of these activities, with varying levels of instructor involvement. For example, an instructor could set up handouts and supplies for all three on separate tables as stations and students would choose what interests them. All three activities are designed to be self-guided, but it is helpful to have a whole-class debriefing at the end about what students learned or observed, in order to draw out the desired illustration of convergence and divergence.

Supplies Needed	Class Time Required	Group Size
Fun with Paper		
Scrap paper (1-3 pieces per student)	20-25 minutes	1-4 students
Scissors (1 pair per student)		
Fun with Fractals		
Straight edge (1 per student)	30-40 minutes	2-4 students
Pencil and eraser (1 per student)		
Fun with Gravity		
Meter stick (1 per group)	45 minutes	3-5 students
Bouncy ball (1 per group)		
Calculators (at least one per group)		
Stopwatches (optional)		

Running the Activity

For each activity, hand out the handouts and supplies, or set up stations containing them. Labeling each table by activity can be helpful if students get to choose their activity. Students should follow the instructions on each handout and answer the questions, discussing with their group as they go. The handout is designed to guide students through the activity, and it is usually helpful to observe what students are doing and ask helpful questions about their progress and thoughts. After students have had time to fully analyze the situation (or run out of time), discuss what they observed and calculated.

Fun with Paper

The handout directs students to cut a piece of $8.50'' \times 11''$ paper into a square, then cut the square into successive halves using two methods, each of which is illustrated on the handout. The handout then has students explore what would happen if they were to continue each process indefinitely. After calculation, students will find that in either case, there would be a finite area of paper in infinitely many pieces, giving a useful concrete illustration of a convergent series. In the first method, the length of paper cut is infinitely long (illustrating a divergent series), while in the second, they would cut a finite distance (illustrating a convergent series). For this activity, one interesting point is that the methods of cutting are ostensibly similar but result in quite different outcomes (finite versus infinite cutting lengths).

Fun with Fractals

After the students draw several iterations of the Koch snowflake, they should begin figuring out lengths and areas as they answer the questions on the handout. They should find that while the added amount of perimeter and area each decrease to zero as the number of iterations tends to infinity, the perimeter tends to infinity while the area tends to a finite amount. This provides both an example of the failure of the converse of the nth term test for series divergence, and somewhat concrete examples of convergent and divergent series. In particular, the students are forced to confront an object with an infinite perimeter and a finite area. This should cause them to be taken aback and to realize that there are situations in which our intuition needs to be verified by mathematics, since most students would consider this impossible.

Fun with Gravity

Prepare a location in which students can drop bouncy balls on a good rebounding surface (such as a tabletop or desktop). Each group will drop the ball several times, take measurements of the rebound height using the meter stick, then model the distance traveled by an ideal ball that rebounds to exactly the same percent of its height each time and continues bouncing in this way forever. They will then apply the projectile equation to analyze the length of time this will take. They should find that the distance and time are both finite (e.g., that their series converge).

For this activity, in addition to the benefit of seeing a concrete example of a convergent series, there is an especially good use of detailed thinking when modeling. For instance, one can discuss the role of the assumption about rebound height, and discuss with students what would happen if they picked a different percent. (Hopefully, they will observe that only a percent less than 100% would be reasonable, and would still result in a convergent series.) Also, it is interesting to discuss how to set up the series given that the ball travels only down for the first distance and both up and down for every subsequent bounce; students could either pull the first trip down out of the series and then reset the initial height, or write the series as though the ball goes both up and down the first time and then subtract off the first distance. You will probably have to lead them into considering how to account for this in their series model.

Suggestions and Pitfalls

Fun with Paper

After several minutes of paper-cutting, it is helpful to remind the students that there is math to be done. Often students begin a competition as to who can cut the smallest pieces of paper; I usually allow this as long as they get to the math in a reasonable amount of time. One of the goals is to have fun, since series is usually a fairly intense chapter. For the mathematical analysis, one can use the actual dimensions of the paper, but it is easier to declare the paper squares 1 unit by 1 unit.

Fun with Fractals

Students nearly always carry a student ID, and this can make a useful straight edge for the level of accuracy needed. It is more difficult to complete this activity with a pen since then one cannot erase the middle line segments before constructing the triangles, so strongly encourage the use of pencils and erasers. Also, it might be helpful to have extra copies of the triangle. In addition, it is helpful to check that students are finding and replacing all line segments with the "mountain" shape; usually they miss the remaining segments of the original triangle.

Fun with Gravity

Students may take a while to get used to estimating rebound height. Help any who are particularly disturbed by inaccuracy to accept the idea of estimating by discussing the nature of mathematical models and perhaps suggesting that they include a statement about what level of accuracy they might be obtaining. For the questions on the time elapsed, students may decide they would rather take data and find an average or pattern that way, instead of using the projectile equation that is given on the handout. It will probably be helpful to push them to collect sufficient data to draw good conclusions in this case. If you wish to include this as part of the activity, you could provide stopwatches.

If you have more ready access to yardsticks than meter sticks, change the units (3 ft rather than 1 m) and the gravitational constant ($s(t) = h - 16t^2$ instead of $s(t) = h - 4.9t^2$).

Fun with Paper – Class Handout

1. Begin with two square pieces of paper. (To make a square piece of paper from a rectangular one, fold the 8.5″ side diagonally down to the 11″ side, and cut off the part that overlaps.)

2. We are going to cut each square of paper in half repeatedly, using two different methods. The first is to cut the paper longways each time. The second is to rotate the paper 90° between each cut, as shown below.

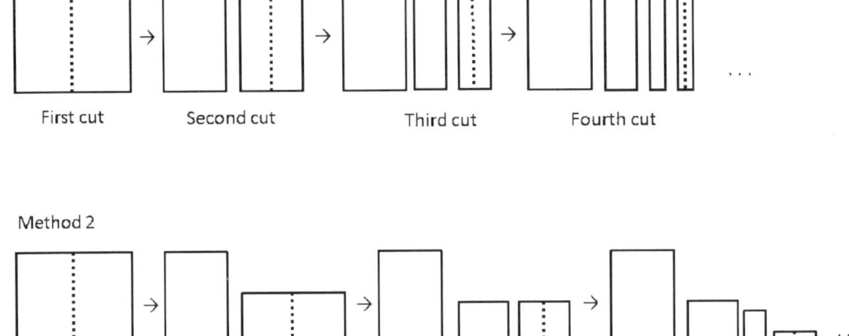

   We will continue each process indefinitely; that is, make infinitely many cuts. (You, of course, will only be able to make a few cuts with your scissors, since eventually the pieces of paper would be too small.)

3. Answer the following questions about each method, *assuming that you have continued the process ad infinitum*. Note that the answers are not all identical for the two methods!

   (a) How many cuts were made?
   Method 1: ___∞___ Method 2:___∞___

   (b) How many pieces of paper will there be?
   Method 1: _______ Method 2:_______

   (c) Assume the original size of the square paper is 1 nice unit by 1 nice unit. How far do the scissors cut in each case? (Another way to ask this question is: How far would the paper extend if you line the pieces up, with all the cut sides along the edge of your table?)
   Method 1: _______ Method 2:_______

   (d) If we don't count time between cuts, how long will the scissors be cutting?
   Method 1: _______ Method 2:_______

   (e) What is the total surface area of all the cut pieces? (Remember that we are assuming "nice units".)
   Method 1: _______ Method 2:_______

   (f) What is the surface area of the nth piece?
   Method 1: _______ Method 2:_______

Triangle for Fun with Fractals

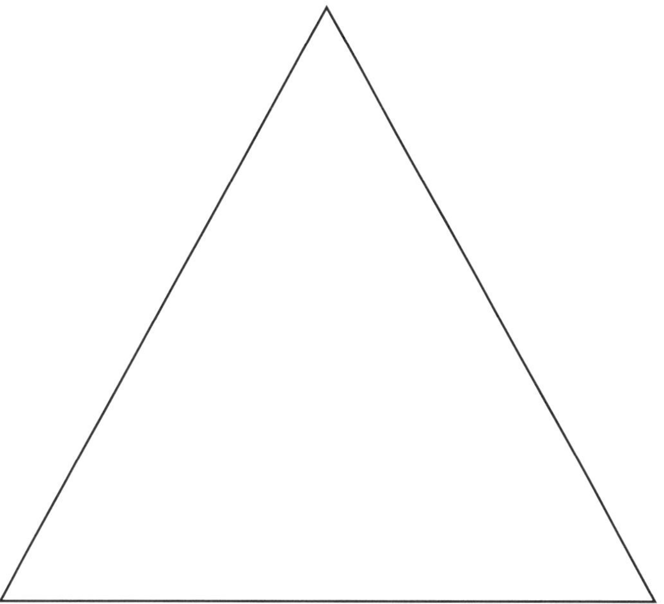

Fun with Fractals – Class Handout

How to make a Koch snowflake:

Step 1. Begin with a drawing of an equilateral triangle.

Step 2. Wherever you see a straight line, draw an equilateral triangle on the middle third of the line segment and erase its base:

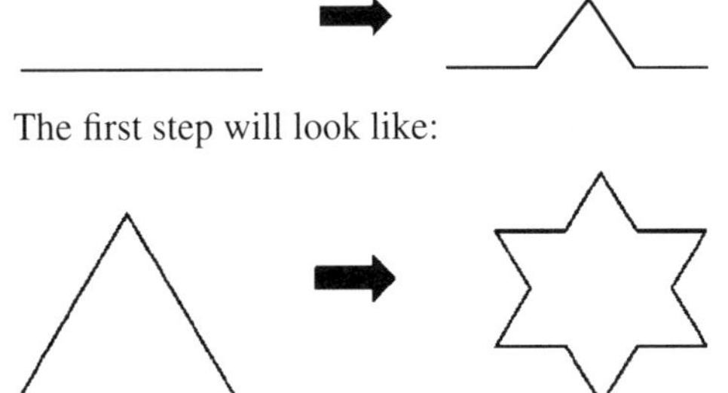

The first step will look like:

Step 3. Repeat Step 2 infinitely many times.

Please draw only an approximation on your paper – stop after enough repetitions to give you an idea of the final outcome. You do not have to construct a formal trisection of your segments, or an exact equilateral triangle, but draw carefully.

Complete the following problems.

1. If the original triangle has side length 1 unit, and the area is A square units, fill in the following chart:

Step (n)	Length of one side (s_n)	Number of sides	Perimeter (p_n)	Area of one new triangle (a_n)	Number of new triangles	Area of snowflake (A_n)
0	1	3	3	n/a	0	A
1	$\frac{1}{3}$	12	4	$\frac{1}{9}A$	3	$A+\frac{1}{3}A$
n						

2. What happens to s_n, the length of each side, as n (the number of iterations) goes to infinity?

3. What happens to p_n, the perimeter, as n goes to infinity?

4. What happens to a_n, the area of one added triangle, as n goes to infinity?

5. What happens to A_n, the total area of the snowflake, as n goes to infinity?

Fun with Gravity – Class Handout

1. Obtain a meter stick and a bouncy ball.

2. Drop the ball *once* from a height of 1 meter, let it bounce three times, and record the height of each of these three bounces.

   This may take some practice and teamwork! Assign each team member a "bounce" to measure – that is, one person watches to see how high the ball goes after the *first* bounce, a second person watches to see how high the ball goes after the *second* bounce, etc.

3. Figure out to what percent of its height the ball rebounds each time.

4. We are going to pretend we live in a perfect world as we model this phenomenon – this is called "making an assumption." Make an assumption about what height the ball would reach on the rebound if dropping from a height of h meters.

   We do still want meaningful results – so this assumption should be based on the data you collected and calculated in steps 2 and 3.

5. Suppose the ball bounced at this rebound percentage infinitely many times. Find the total distance travelled by the ball, if it is dropped from a height of 10 meters. (Hint: Write an infinite series.)

6. From physics, the height of the ball at time t if it is dropped from a height of h meters with an initial velocity of 0 meters per second is $s(t) = h - 4.9t^2$. How long does each of the first three up-and-down bounces take the ball, if it is dropped from a height of 10 meters?

7. Find the total amount of time that the ball would bounce. (Hint: Write an infinite series. Look for a pattern in your answers in the previous step, or do some algebra.)

Chapter 4

Multivariable Calculus

4.1 Physically Creating Three Dimensional Graphs

Julie Barnes, Western Carolina University

Concepts Taught: graphing three dimensional functions

Activity Overview

Unless your students are all talented artists, it can be difficult for them to sketch the graphs of functions of two variables that are typically found in multivariable calculus. There are many software packages that help some students; however, other students may not immediately see the connection between the two dimensional images on the computer screen and the three dimensional nature of the graphs. In this activity, students are able to walk around a three dimensional coordinate system made of yarn, create various graphs either with their bodies or with yarn, and physically experience what the various graphs tell them. In addition, they directly experience the ideas of cross sections by either being a cross section themselves or holding the yarn that represents it.

Supplies Needed	Class Time Required	Group Size
Yarn Scissors Pre-assembled coordinate system *	30-40 minutes	Class demonstration

*Before doing the activity the first time, you will need to assemble the three dimensional coordinate system. You need three pieces of yarn, each of a different color; at least two of these pieces should be 12-15 feet long, while the third only needs to be about 8 feet long. Lay the two large pieces of yarn on the floor or a table and tie them together in the middle so that now four pieces of yarn extend from the center; these represent the xy-coordinate system. The other piece of yarn should be tied to the first two at the location of the knot just created, but for this new piece only about 2 feet should extend on one side of the knot; this piece of yarn represents the z-axis. It is convenient to tie a small weighted object to each end of the x-axis and y-axis to help with storage. Wooden letters, like Ns for the negative ends of both the x-axis and y-axis and Hs or 8s for the positive ends of the x-axis and y-axis, are nice because the shape makes it easy to wrap yarn around it. You can also use pieces of masking tape or packaging labels to mark 1 foot units along the yarn. See Figure 4.1a for a photograph of how the pre-assembled axes should look.

This activity requires an open space of at least 10 ft $\times$ 10 ft.

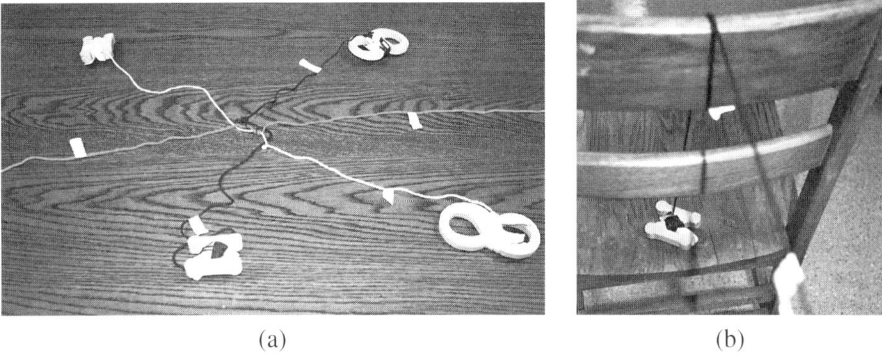

(a) (b)

Figure 4.1: (a) Assembled coordinate system before taking it to class. (b) One way to set up the coordinate system is to loop the ends of the yarn over the back of a chair and pull the string tight. There is no need to tie any knots.

Running the Activity

Bring the pre-assembled coordinate system to class, along with extra yarn, scissors, and copies of the handout. At the beginning of class, have students help you set up the coordinate system. Tape the z-axis to the ceiling, or if the ceiling is too high, have a tall student hold the positive z-axis in the air. The x-axis and y-axis can each be stretched out and attached to chairs. If you use the weighted objects, there is no need to actually tie knots at the chairs; just loop the weighted objects around the back of the chair, and the weight of the objects usually holds the yarn in place as in Figure 4.1b. It is best to have the x-axis and y-axis a couple of feet above the floor.

Provide each student with a copy of the handout, which contains a list of functions of x and y. Students can take notes on the handout as the class works through the examples listed. Tell the class that their challenge is to use yarn and their bodies to represent each function on the handout. It is up to them to decide how they will accomplish this. To plot the point $(3, 2, 1)$, students typically walk to the location and place a finger where the point is. To represent a line, like $x = y = z$, students tend to have a couple of volunteers hold yarn as shown in Figure 4.2a. When representing a plane or a cylinder, like $x = y$ or $x^2+y^2 = 0$, students often use their bodies, as shown in Figures 4.2b and 4.2c. For a parabolic cylinder, students might choose to use their arms to represent the graph, as seen in Figure 4.2d. The nice thing about this representation is that each student is essentially one of the vertical cross sections, each with a fixed y value. For $z = 3$, students tend to just describe what it is and why it would be difficult to have people represent it. However, sometimes students have held notebooks up horizontally three feet above the xy-plane. Similarly, for describing the set of points one unit from $(0, 0, 0)$, students typically just describe the sphere and maybe hold their hands around where it would be, like a potter molding clay.

The function $z = x^2 + y^2$ is harder to represent with just yarn or arms and will take roughly half of the time allotted to the entire activity. It is well worth that time because it is a great way to connect the first part of this activity to cross sections, contours, and any computer images of functions they may see later in the course. You may want to help direct this one. The best way for students to graph this function physically is by using yarn to graph cross sections. Place students around the axes system so that any points you want to include are reachable by at least one student. Some good points to begin with are $(0, 0, 0), (1, 1, 2), (1, -1, 2), (-1, 1, 2), (-1, -1, 2), (1, 2, 5)$; if you have enough students, include more points as desired. Then use string to create the vertical cross sections at $x = 0$, $x = 1$, $x = -1$, $y = 0$, $y = 1$, $y = -1$, and any other cross sections your students are able to reach as in Figure 4.2e. If possible, use two different colors of yarn for cross sections parallel to the xz-plane and cross sections parallel to the yz-plane. While doing this, ask students what this yarn represents, what shape it is, and how it relates to the equation of the function they are plotting. Once that is plotted, it is nice if you can use a third color to plot horizontal cross sections for $z = 1$ and $z = 2$ as in Figure 4.2f. Ask students why you aren't plotting $z = -1$ and why $z = 0$ is not a circle like the first two were. You can also point out that if those circles from $z = 1$ and $z = 2$ were all moved down to the $x = y$ plane, they become contours or level curves.

At the end, have the class summarize what they have observed while physically graphing these functions.

Suggestions and Pitfalls

When you are done, roll the yarn onto the objects tied to the ends. The axes store well this way.

The last problem about $z = x^2 + y^2$ will require 5-20 students depending on how many cross sections you plot and whether students are holding one point each or two. You should reserve taller students for the points further from the origin.

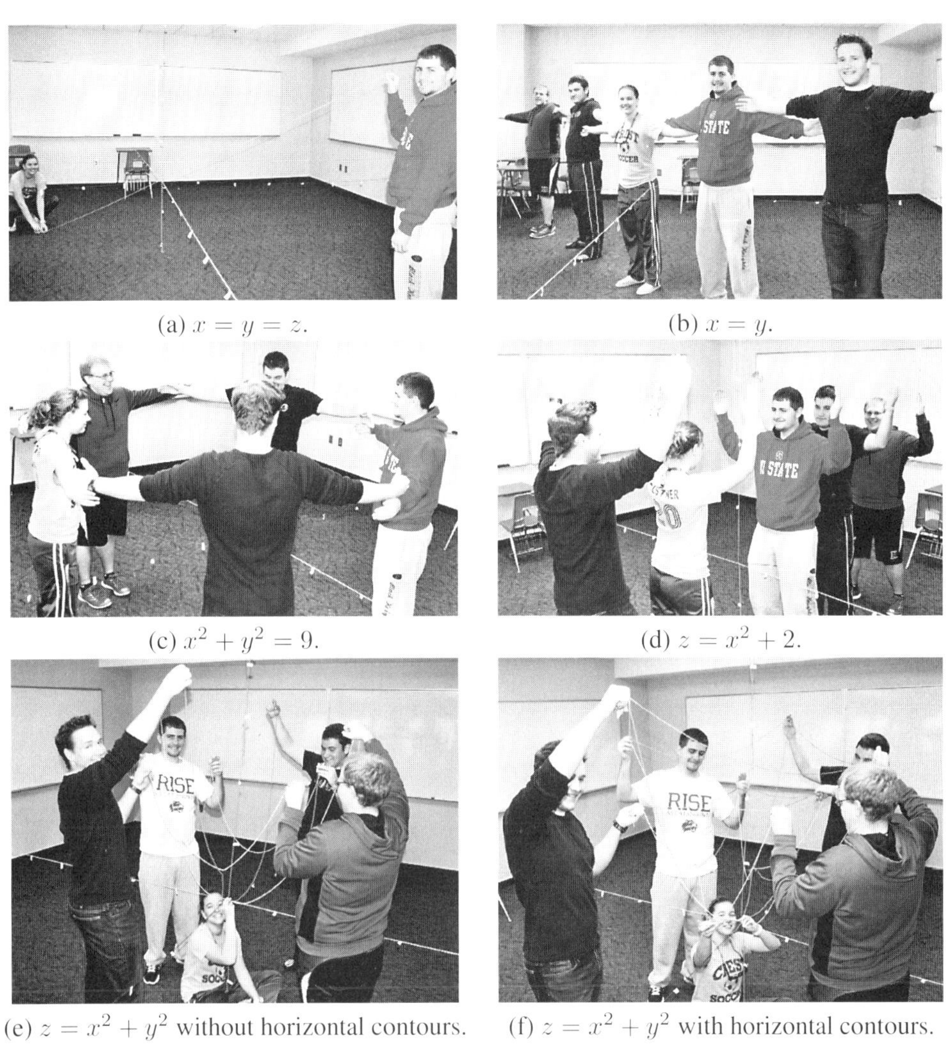

(a) $x = y = z$.

(b) $x = y$.

(c) $x^2 + y^2 = 9$.

(d) $z = x^2 + 2$.

(e) $z = x^2 + y^2$ without horizontal contours.

(f) $z = x^2 + y^2$ with horizontal contours.

Figure 4.2: Students representing three dimensional graphs.

Physically Creating Three Dimensional Graphs – Class Handout

As you and your classmates create human and yarn versions of the graphs for each of the following functions, briefly record what the graph looks like.

1. Plot the point $(3, 2, 1)$.

2. Graph $x = y = z$.

3. Graph $x = y$.

4. Graph $x^2 + y^2 = 9$.

5. Graph $z = x^2$.

6. Graph $z = 3$.

7. Describe the set of points that are one unit from $(0, 0, 0)$.

8. Graph $z = x^2 + y^2$.

4.2 Building Functions of Two Variables with Cookies

Jessica Libertini, Virginia Military Institute

Concepts Taught: Concepts taught: multivariable functions, partial derivatives, Riemann sums, volume approximations, contour lines

Activity Overview

When first learning about functions of one variable, students are often asked to evaluate a function at several points, plot those points, and sketch the graph to help them understand the shape and behavior of that function. As we move to functions of two variables, paper sketches become problematic, requiring us to look at cross-sections or develop strong perspective drawing skills. While software packages such as Mathematica allow students to visualize functions of two variables, they cannot replace the intuition students develop by plotting points in space and generating their own graphs of multivariable functions. In this activity, students use cookies to generate three dimensional graphs of functions of two variables. This activity also provides a useful foundation for a variety of topics such as partial derivatives, multivariable integration, and contour lines as explained in the Extensions Section on page 117

Supplies Needed (per group)	Class Time Required	Group Size
One large package of cookies* Cookie-sized coordinate system**	20-30 minutes for activity 10 minutes per extension	3-5 students

*Each group will need between 20-60 stackable cookies or crackers to make their functions, so if using packages that have less than 60 cookies, encourage groups to share. Alternatively, this activity can be done using coins or other stackable objects, as shown in Figure 4.5.

**Based on the size of the cookies, you will need to make a handout for each group with an appropriately sized coordinate system on it; in other words, the unit length of the grid is the diameter (or the major axis, for noncircular cookies) of a cookie. Since each group will be graphing a different function over a different domain, it is best to have the students label the axes and origin.

Figure 4.3: Three dimensional graph of a function of two variables using cookies.

Running the Activity

Divide the class into groups of 3-5 students. You may wish to have them rearrange their desks so they can share a work space. Provide each group with a box or bag of cookies, a cookie-sized coordinate system, and a copy of the handout. Assign each group a domain and a function of two variables whose value is positive over the prescribed domain. Each group of students should then calculate, for each pair of coordinates on the grid, the number of cookies they need to stack in order to build a three dimensional representation of their function.

Problems 1 and 2 on the handout guide groups through the creation of a three dimensional cookie-graph of their function, as shown in Figure 4.3. Problems 3 and 4 address the limitations of using cookies to plot functions. Problem 5 asks students to discuss what other groups' graphs might look like based on equations. Lastly, Problem 6 has them visit other groups to see how the cookie graphs compare with their answers to Problem 5. Be sure to allow time for them to do the gallery walk to see the work of their classmates.

Suggestions and Pitfalls

In order to see a nice portion of the graph while minimizing time, take advantage of symmetry. For the sample functions given in the handout, it is best to make a 5×5 grid centered about the origin. OREO® cookies are approximately $1.75''$ in diameter, so you can fit four across a sheet of paper, but not quite five. Generic cookies are often a little smaller, allowing you to fit a row of five cookies with boxes of $1.7''$ or smaller.

Be sure that you've calculated the number of cookies required to graph each function and that each group has access to enough cookies to complete their graph. Sandwich cookies, such as OREO cookies, give students the opportunity to be creative in representing non-integer values as seen in Figure 4.4a. You may wish to have them present their graph side by side with a computer-based representation, such as a three dimensional plot or a contour plot, as shown in Figure 4.4.

(a) (b)

Figure 4.4: Three dimensional cookie graphs displayed with corresponding Mathematica contour plots.

Figure 4.5: Three dimensional graph of a function of two variables using coins.

Extensions

This activity can be used simply to introduce graphs of functions of two variables; however, it can also be used to support several multivariable calculus concepts, as described below.

Partial Derivatives and Directional Derivatives. Due to the inherently discrete nature of the cookie graph, students can approximate directional derivatives and partial derivatives by comparing the heights of neighboring stacks of cookies. For example, you might ask each group to use their cookie graph to approximate the partial derivative at $(0, 1)$ in the y-direction.

Volume Approximations. When making a graph in three-space, it would be wonderful if we could hang points, or cookies, in midair; however, in this exercise, we represent the height of the function at a point by building a stack of cookies. The lower cookies, while not actually part of the graph of the function, literally play a supporting role (holding up the upper cookie); they also represent an approximation of the volume beneath the surface. Therefore have your students approximate the volume captured between their surface and the xy-plane by totaling the number of cookies used in their graph.

Contour Plots and Contour Lines. For this add-on activity, you should have markers or crayons available. Have your students replace each stack of cookies with a color-coded marking, e.g., all blocks that contained stacks of 3 cookies get colored red while all blocks containing stacks of two cookies get colored orange, etc. By having your students connect similarly colored blocks, they can explore contour lines. Ask your students what it might mean for two adjacent blocks to have the same color.

Building Functions of Two Variables with Cookies – Class Handou2t

Your group will be assigned one of the functions and domains from the list below.

Function	Domain
$f(x,y) = \|x\| + \|y\|$	$D = \{(x,y) : -2 \leq x \leq 2, -2 \leq y \leq 2\}$
$f(x,y) = \|xy\|$	$D = \{(x,y) : -2 \leq x \leq 2, -2 \leq y \leq 2\}$
$f(x,y) = \|4 - x^2 - y^2\|$	$D = \{(x,y) : -2 \leq x \leq 2, -2 \leq y \leq 2\}$
$f(x,y) = (x-y)^2$	$D = \{(x,y) : -2 \leq x \leq 2, -2 \leq y \leq 2\}$
$f(x,y) = \|x^2 - y^2\|$	$D = \{(x,y) : -2 \leq x \leq 2, -2 \leq y \leq 2\}$
$f(x,y) = \|\sin(x) + \sin(y)\|$	$D = \{(x,y) : -\pi \leq x \leq \pi, -\pi \leq y \leq \pi\}$

1. Label each block on your grid with the ordered pair it represents.
2. Calculate your function's value using the x and y values for each ordered pair, and build a cookie stack of that height in the corresponding box.
3. You have only plotted a limited number of points. Describe what your shape might look like if your domain were larger. Will your function ever be negative, and if so, where?
4. Your cookies are much larger than points, making the graph *chunky*. What might your function look like if we had cookies with a much smaller diameter?
5. Look at the other functions above, and discuss with your group members what the graph might look like. Write a short description or make an informal sketch for each in the space below.
   (a) $f(x,y) = |x| + |y|$
   (b) $f(x,y) = |xy|$
   (c) $f(x,y) = |4 - x^2 - y^2|$
   (d) $f(x,y) = (x-y)^2$
   (e) $f(x,y) = |x^2 - y^2|$
   (f) $f(x,y) = |\sin(x) + \sin(y)|$
6. Visit the other groups to see their graphs. Do the graphs look the way you expected? If not, what was different?

4.3 Exploring Contours in the Physical World

Elin Farnell, Colorado State University
Shawn Farnell, Colorado State University

Concepts Taught: contour diagrams, directional derivatives, gradient vectors

Activity Overview

This activity aims to develop student understanding of contour diagrams within a multivariable calculus course. Contour diagrams are frequently used to introduce and discuss many of the constructs that appear in a multivariable calculus course, such as directional derivatives, linear functions, and gradient vectors. The following activity is highly useful for developing student visualization skills as well as for foreshadowing many important multivariable calculus concepts. In the activity, students work in pairs to lay down string approximating contours on the ground in a nearby outside location* and answer related questions.

Supplies Needed (per group)	Class Time Required	Group Size
1 piece string or ribbon, approximately 30' long	30-50 minutes depending on amount of discussion	2 students

*It is recommended that the outside location have features that will lead to variation in contours. For example, trees create good examples of ridges as well as qualitatively different behavior from nearby ground as a result of roots and planting practices. It is also helpful to have a hillside and an inclined sidewalk.

(a) (b) (c)

Figure 4.6: (a) An example of an ouside location. (b) A side view of contours around a tree trunk. (c) An aerial view of contours placed around a tree trunk.

Running the Activity

Prior to starting the activity, briefly introduce contour diagrams. Then, divide the class into pairs, and give one string or ribbon to each student pair. We will hereafter refer to the use of string, though both string and ribbon work well for this activity. Head outside (Figure 4.6a) and instruct each pair to place their string on the ground to create a contour that satisfies the following conditions.

1. String contours should be placed so that the vertical distance between any two adjacent contours is 6 inches; if the terrain is particularly shallow or steep, the instructor may decide in advance to use a different fixed elevation, typically between two inches and two feet.

2. The placed string should cross a sidewalk. If there is no sidewalk in the space available, we recommend placing something like a large board or cardboard on the ground that the string contours must cross. This second requirement works well to provide for later discussion of linear functions and local linearity.

After students have finished placing their string contours (Figures 4.6b and 4.6c), provide them with a handout, and have them complete the first problem which asks them to sketch the contours created. Then, use the problems on the handout as an outline while engaging the students in a whole-class discussion. Encourage students to use the handout to organize their notes. This phase of the activity provides an opportunity for students to note and discuss general behavior such as the relationship of the closeness of contours on their two dimensional drawing as it relates to the relative steepness of the hillside. Problems 2, 3, 4, and 5 help students understand how the shape of the terrain is related to the position of the contour lines.

Before talking about Problem 6, instruct students to stand on a contour at any point. Then ask them to face in the direction in which their elevation would increase the fastest if they took one small step. Have students discuss their chosen direction as it relates to the nearby placed contours. Have the students turn 90° and ask what is special about the resulting direction; repeat this for two more 90° turns. There are several important concepts to emphasize at this stage about instantaneous rates of change: the greatest rate of increase at any point is perpendicular to the contour through that point, the greatest rate of decrease is in the opposite direction (and is also therefore perpendicular to the contour), the greatest rate of increase or decrease is a local property, and moving in the direction of the contour results in a rate of change of zero.

As you look at Problem 7, be prepared to discuss future topics that you would like to foreshadow. For example, ask students to estimate the instantaneous rate of change in a chosen direction to introduce the idea of a directional derivative. This is a good point at which to discuss what information is needed: change in height (most easily estimated by using the closest contour) and change in horizontal distance (as measured in an imagined xy-plane).

Finally, Problem 8 sums up the activity by asking students to discuss the pros and cons of contour diagrams. Important contributions to the discussion include the ability to view the qualitative behavior of a function over a region, the ease with which slope, critical points, heights, etc., may be obtained, and the two dimensional nature of a contour diagram versus the three dimensional nature of the graph of a surface.

Suggestions and Pitfalls

Give students time to lay down their string and encourage them to walk along it. Remind students that they should feel no change in elevation as they walk along their string if it is truly a well-placed contour. It is also beneficial to view the string from afar and to compare it with contours placed by other student pairs.

Use the chosen location for the activity to determine the number of strings that will best capture the behavior of the landscape. If the class size is small, it may be desirable to give multiple strings to each student pair. If the class is very large, group sizes can be increased to 3-4 students.

If there is a nearby building with high floors and an accessible window, it may be worthwhile to take a picture from above to get a good representation of the student-created contour diagram for later discussion.

Obtaining perfect contours can be difficult; it is far more important to generate useful discussion about general contour behavior and to initiate critical thinking about relevant multivariable calculus topics.

It may be helpful to have students use modeling clay to construct a surface with a contour that has multiple disjoint curves or intersections as a homework problem or follow-up in-class activity.

Exploring Contours in the Physical World – Class Handout

1. Sketch a contour diagram on the back of this handout that incorporates the contours placed by the class.

2. Note that regions in your contour diagram in which contours are close to each other correspond to relatively steep regions in the landscape. Give an intuitive explanation for this relationship.

3. What is the behavior of the contours on planar regions, such as on a section of a sidewalk?

4. How do contours behave around peaks and valleys? Is it possible to tell from a contour diagram alone what happens inside the innermost contour?

5. Is it possible for two strings to cross? More specifically, some possibilities to consider are the following. Can a contour consist of multiple disjoint curves? Can a contour consist of intersecting curves? Can contours corresponding to different heights intersect? Explain.

6. What is the relationship between the direction of greatest rate of increase at a point and the contour through that point? What can you say about the direction of greatest rate of decrease at a point?

7. Describe how to estimate the instantaneous rate of change in a specific direction.

8. Discuss some pros and cons of using a contour diagram vs. a three dimensional graph of a function.

4.4 Matching Photographs with Contour Lines

Julie Barnes, Western Carolina University

Concepts Taught: contour lines, three dimensional functions, spatial reasoning

Activity Overview

By the time students arrive in multivariable calculus, they are typically very comfortable with studying two dimensional graphs of functions of one variable. However, they are typically not comfortable working with two dimensional representations of three dimensional graphs. In this activity, students match a collection of photographs with their corresponding topographical maps, i.e., contour plots. By comparing photographs of real scenery to maps of these scenes, two dimensional representations of three dimensional surfaces can become more concrete. In addition, this gives students some practice looking at surfaces from different directions and using different representations.

Supplies Needed	Class Time Required	Group Size
None	10 minutes	2-3 students

Running the Activity

Provide each group with a copy of both pages of the handout. Have students work in groups to pair the photographs with their matching topographical maps.

While they are working, walk around to see how students are doing. Some students will find this exercise difficult and will need some assistance. Some useful hints are given below.

> Where are the cliffs and mountains? How will contour lines look there?
>
> Photo A was taken from a beach. What is the elevation there?
>
> Photo B and Photo C were both taken from the edge of a steep cliff. What kinds of contour lines do you expect from that?
>
> The hills seen in Photo C are not as steep as the cliffs in the other two photos. How is that depicted in the contours seen from the viewing point?

The main goal of the activity is to have students think about and discuss how different features in the photographs appear in the contour maps and vice versa. Actually making the correct matches is less important, but students will want to know if they are right. The correct matching for the provided photographs is Photo A and Map 2, Photo B and Map 3, and Photo C and Map 1.

Suggestions and Pitfalls

You can use the photographs and topographical maps provided, or create your own from photographs taken locally. The topographical maps here are from the USGS (United States Geological Survey) and are available online.

If using your own photographs, color images tend to work better than black and white.

Part I: Matching Photographs with Contour Lines – Class Handout

Match the photographs in Figure 1 with the topographical maps provided in Figure 2.

Photo A

Photo B

Photo C

Figure 1: Photographs of various scenes in Hawaii.

Part II: Matching Photographs with Contour Lines – Class Handout

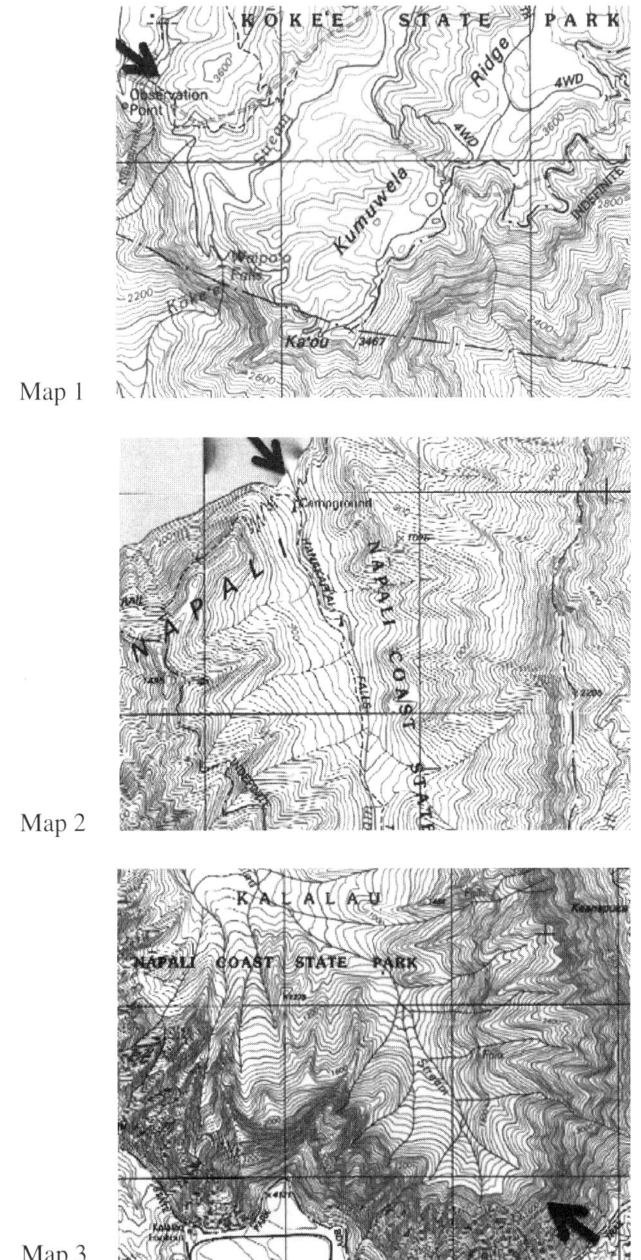

Figure 2: Topographical maps that correspond to the photographs in Figure 1. Each arrow represents where the photographer was standing and the direction he/she was facing.

4.5 The Gold Mine: Tangent Plane Approximation Using Tangible Surfaces

Brian Fisher, Lubbock Christian University

Jason Samuels, City University of New York - BMCC

Aaron Wangberg, Winona State University

Eric Weber, Oregon State University

Concepts Taught: linear approximation, tangent plane, directional derivatives

Activity Overview

How could you estimate the density of buried gold in the ground on your neighbor's property without trespassing? Based on a gold-mining scenario, this activity, which uses a surface like the ones shown in Figures 4.7 and 4.8, is designed to let small groups of students explore the properties of tangent planes before they are introduced in lecture. It assumes students are familiar with partial and mixed partial derivatives. Students develop the notion of a tangent plane approximation by taking local measurements on their surfaces. Then they identify local quantities that are identical on both the surface and the local tangent plane and discuss how other quantities impact the accuracy of the tangent plane approximation. This material is based upon work supported by the National Science Foundation under Grant Number DUE-1246094.

Supplies Needed (per group)	Class Time Required	Group Size
Pre-made surface* Coordinate grid Dry erase markers Tangent plane** Slope-meter*** Ruler	25 minutes	3 students

*The surface in Figure 4.7a has a clear, dry erasable finish, but any wavy, stable object will suffice, such as the form in Figure 4.7b which is made of plaster of paris. Mark a blue dot on the surface at a point where the surface is concave down and the gradient is non-zero. Neither the surfaces or the location of the blue dots needs to be standardized for the class.
**Cut a 3 inch by 3 inch tangent plane from cardboard, thick transparencies, or similarly rigid yet flexible and thin material. See Figure 4.8.
***The students will need to measure the slope of a line tangent to their surfaces, so a simple tool like that shown in Figure 4.7a and 4.7b can be constructed using two dowels,

©2014 Brian Fisher, Jason Samuels, Aaron Wangberg, Eric Weber

a hinge, and a bubble level. Students then use the ruler to measure both the rise and run to calculate the slope. Alternatively, students could use one of several available free angle-measuring inclinometer smartphone apps, but then they would need to convert from the angle to the slope.

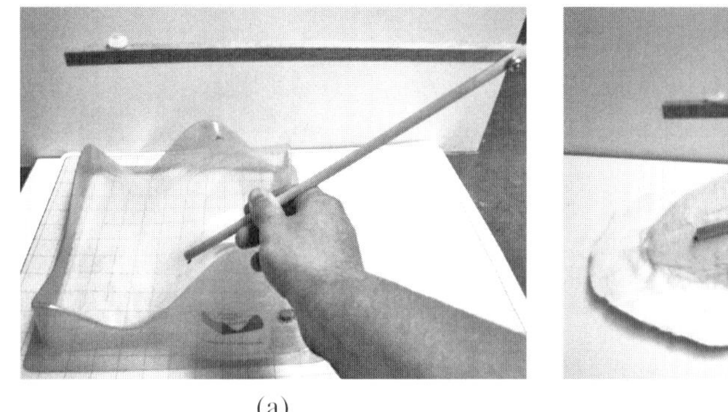

(a)

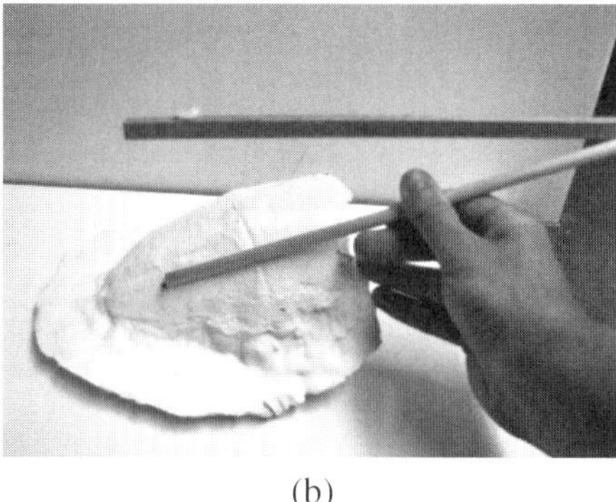

(b)

Figure 4.7: (a) A slope-meter estimates the directional derivative on a tangible surfaces made of plastic or (b) plaster of paris.

Running the Activity

To begin the activity, divide students into groups of three. Distribute the handout and the first four supply items. The handout is divided into sections titled **On Your Mark**, **Get Set**, **Go**, and **Challenge**. Withhold the slope-meter and ruler until ready for **Get Set**.

On Your Mark: For this section of the handout, give students 5-10 minutes to generate their answers. Encourage students to draw the quantities on the surface and tangent plane. First order derivatives are the same on both the surface and tangent plane, but all second order derivatives vanish on the plane. Before proceeding, have the groups present their answers. Ask students whether gradient vectors, directional derivatives, or level curves will be the same for the surface and the plane at the blue dot (they are), or whether their answers depend on the location of the origin (they don't).

Prior to **Get Set**, show how to use the slope-meter to estimate slope at a point. Draw the graph of $y = 5 - x^2$ on the board and estimate the slope at $x = -2$. The round dowel is placed tangent to the graph. The square dowel is adjusted to be horizontal; the bubble level ensures the square dowel is horizontal. Laying the slope-meter on the grid, it is possible to count boxes to measure the rise (perpendicular to the square dowel) and the run (along the square dowel), then divide to get the derivative. Distribute the slope-meters and rulers.

Get Set: At any location, the surface height represents the density, ρ, of buried gold (in g/km^3) beneath the ground. Students should estimate the surface's height using the ruler and its derivatives with the slope-meters. Help groups use the slope-meters. We let students define north and east in their choice of coordinates. The derivatives should have units of g/km^4. Unit analysis is helpful for **Go**.

Go: Students need to combine the quantities from **Get Set** with the displacements. Mine A has a zero displacement in the east-west direction, while Mine C intentionally has a displacement in the south, not north, direction. Remind students they can only use information measured at the blue dot to estimate the density of gold at each mine. Remember: No Trespassing – this is gold country!

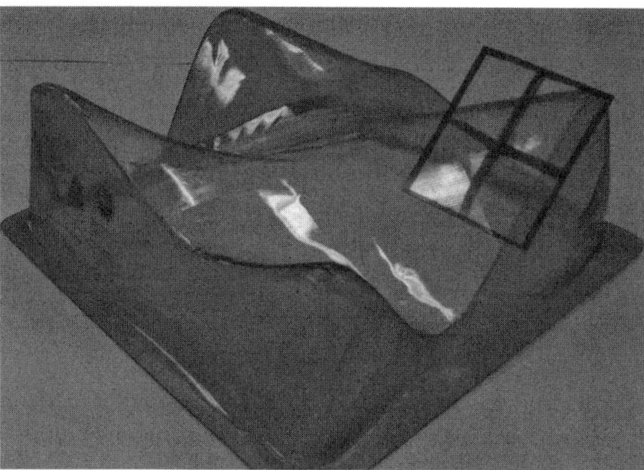

Figure 4.8: A tangent plane represents the linear approximation to a multivariable function.

Challenge: Students should be able to write the equation of the tangent plane approximation using their coordinates and partial derivatives measured at the blue dot.

Wrap-up Discussion: Conduct a whole-class wrap-up discussion once most groups have finished the **Go** section. Have students demonstrate their estimations from **Go**, asking them to identify whether quantities exist on the domain, surface, or tangent plane. Ask groups which mine they should buy, as their tangent plane estimations often conflict with the actual values given by the surface. This is a great opportunity to discuss how second order derivatives (listed in **On your Mark**) impact error.

Suggestions and Pitfalls

One feature of the activity is that it does not specify an origin or a coordinate system. Students may need to be reassured that they are free to choose the coordinate system and coordinate directions. The answers for **Get Set** and **Go** are independent of the origin's location. Note that estimates of the density of buried gold depend only on the displacement and rate of change in the given directions.

Units and quantities are particularly important in **Go**. The coordinate grid is measured in kilometers, as are all displacements in the domain, regardless of direction. The surface gives the density of gold in terms of mass per volume; level curves have units of g/km^3. Directional derivatives and gradient vectors have units of g/km^4. This work was supported by NSF Grant DUE 1246094. Any opinions, findings, an conclusions or recommendations expressed in this volume are those of the author(s) and do not necessarily reflect those of the National Science Foundation.

Reference

1. A. Wangberg, *Raising Calculus to the Surface: Instructor's Guide and Activty Notes*, self-published (2014).

The Gold Mine – Class Handout

On Your Mark: The plastic square represents a plane. Press it against the surface at the blue dot. Which quantities below are the same on the surface and on the plane? Which are different?

Surface f at the blue dot	$\frac{\partial f}{\partial x}$	$\frac{\partial f}{\partial y}$	$\frac{\partial^2 f}{\partial x^2}$	$\frac{\partial^2 f}{\partial y^2}$	$\frac{\partial^2 f}{\partial x \partial y}$
Plane P at the blue dot	$\frac{\partial P}{\partial x}$	$\frac{\partial P}{\partial y}$	$\frac{\partial^2 P}{\partial x^2}$	$\frac{\partial^2 P}{\partial y^2}$	$\frac{\partial^2 P}{\partial x \partial y}$

Get Set: The surface represents the density ρ of gold (in g/km^3) beneath the ground. You own a small mine located at the blue dot. Estimate the density of gold at your mine, and measure how the density of gold changes in the north and east directions. Use appropriate notation and include units.

(Note: 1 vertical inch = 1 g/km^3 of gold; 1 horizontal inch = 1 kilometer.)

Go: You want to buy one of three mines that are for sale; their locations (relative to your mine) are given below. Estimate the density of gold at each mine using only your previous measurements.

Mine A	Mine B	Mine C
1.2 km north	1.2 km north	3.4 km south
	0.8 km east	1.7 km east

Challenge: Develop a general formula to estimate the density of gold for a mine located at a point (a, b).

©2014 Brian Fisher, Jason Samuels, Aaron Wangberg, Eric Weber

4.6 Visualizing Second Order Partials on a Football

Julie Barnes, Western Carolina University

Concepts Taught: second order partial derivatives, mixed partials

Activity Overview

When students reach multivariable calculus, they are typically comfortable with relating a first derivative to slope and a second derivative to concavity. For a function $f(x, y)$ on two variables, these notions translate easily to comparing f_x and f_y to slope in the x and y directions, as well as comparing f_{xx} and f_{yy} to concavity in the x and y directions. However, students are typically unable to determine a physical interpretation for the mixed partials f_{xy} and f_{yx}. In this activity, students look at a physical model made from a foam football and toothpicks that builds on their intuition of f_x, f_{xx}, f_y, and f_{yy} to better understand the meaning of f_{xy}.

Supplies Needed (per group)	Class Time Required	Group Size
Prepared foam football still covered in original plastic wrap, with 7 toothpicks* 1 extra toothpick	10 minutes	2-4 students

*Before class, prepare foam footballs as depicted in Figure 4.9. Start with a toy foam football that still has original plastic wrap on it. Draw an xy-coordinate system on a piece of paper and place a football on it with the pointed ends of the football along the y-axis. As you look down on the football from above, you see the top half of the football, i.e., the part of the football that is farthest above the piece of paper. Let the height of this surface above the paper be $f(x, y)$. See Figure 4.9a for a diagram of the football's orientation. Use a permanent marker to draw on the plastic covered football directly above a line that is parallel to the y-axis and not the highest ridge of the surface $f(x, y)$ because the mixed partial along the highest ridge is just 0. The line you need to draw corresponds to the vertical line in the lower-left corner in Figure 4.9a as well as on the football in Figure 4.9b. Use a toothpick to poke seven holes in the plastic at approximately $1/4$ inch increments along the curve you drew. Then at each hole, slide in a toothpick so that it is just inside the plastic, lies tangent to the surface, and is parallel with the x-axis. The toothpicks correspond to the horizontal lines in Figure 4.9a with the dotted portion under the plastic; the actual toothpicks can be seen in Figure 4.9b. Footballs with plastic wrap can be found at most discount stores.

Running the Activity

Provide each group of students with a prepared football, a toothpick not on the football, and a handout. Ask them to work through the handout and be ready to provide assistance as they work.

The first problem of the handout asks about partials in the x-direction. Students can observe f_x by holding their extra toothpick on the surface so that it is parallel with the x-axis. The slope of the toothpick in the xz-plane is the value of f_x. As students move the toothpick in the x-direction while remaining tangent to the football, the slope of the toothpick will decrease, indicating that f_{xx} is negative, because the ball is concave down in that direction. This corresponds to what students already know from single variable calculus. The second problem in the handout discusses what happens in the y-direction, and the results are almost identical since the football is concave down in that direction as well.

Once students are comfortable with visualizing f_x, f_{xx}, f_y, and f_{yy} on the football, they are ready to move to the last problem. Problem 3 asks students to look at the model to see how the toothpicks' slopes in the x-direction change as the point of tangency of the toothpick is moved in the y-direction. The sign of f_{xy} for the model we created is negative because the slopes of the toothpicks are positive and become less steep as the base points get closer to the x-axis as seen in Figure 4.9c.

Suggestions and Pitfalls

Sometimes the toothpicks move a bit when you take the model(s) to class. Be sure to have the students fix any toothpicks that are not perpendicular to the line drawn on the football.

Even though the plastic covering on the football appears to be fragile, it is stronger than it looks. I have used some of the same footballs for ten years, and the plastic, for the most part, is still fine. However, if it does rip, there are two easy solutions. Either rotate the football around and draw a new curve on the other side of it where the plastic is still intact, or use cellophane tape to fix any rips and continue to use the same location.

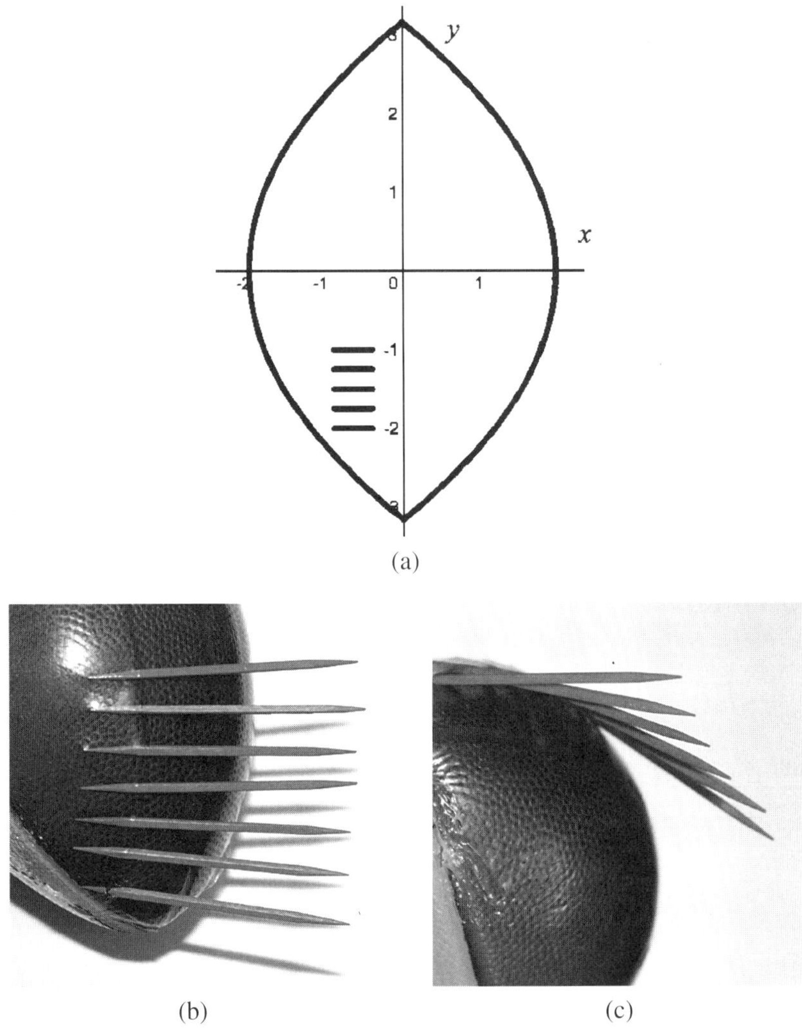

Figure 4.9: (a) Diagram of the football and location of toothpicks. The dotted portion of the toothpicks is under the plastic wrap. (b) Football with toothpicks attached. This aerial view shows that all the toothpicks are parallel to the x-axis as depicted in (a). (c) The side view of the football showing how the slopes of the toothpicks change as the base point moves in the y-direction.

Visualizing Second Order Partials on a Football – Class Handout

Draw an xy-coordinate system on a piece of paper, and place a football model on it with the pointed ends of the football along the y-axis as seen in the diagram on the right. As you look down on the football from above, you see the top half of the football, i.e., the part of the football that is farthest above the piece of paper. Make sure that the toothpicks are completely visible from the aerial view in the lower-left corner, as seen in the diagram; if they are not, rotate the football until they are visible. Let the height of this surface visible from above be $f(x, y)$. Then complete the problems below.

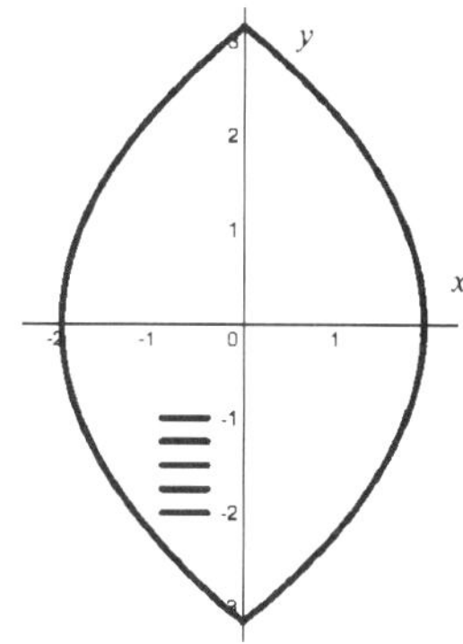

Football Orientation

1. Pick any point on the football. Hold a toothpick on the surface so that it is parallel with the x-axis.
   (a) Is f_x positive or negative? How can you tell? Does your answer differ based on the location of the point you picked?
   (b) Slide the toothpick to the right. As you move the toothpick, what happens to f_x? Does it increase or decrease? What does this tell you about f_{xx}?
   (c) While you moved the toothpick, you traced out a curve along $f(x, y)$. Is that curve concave up or down in the xz-plane? How is concavity related to the sign of f_{xx}?

2. Pick any point on the football. Hold a toothpick on the surface so that it is parallel with the y-axis.
   (a) Is f_y positive or negative? How can you tell? Does your answer differ based on the location of the point you picked?
   (b) Slide the toothpick in the positive y-direction. As you move the toothpick, what happens to f_y? Does it increase or decrease? What does this tell you about f_{yy}?
   (c) While you moved the toothpick, you traced out a curve along $f(x, y)$. Is that curve concave up or down in the yz-plane? How is concavity related to the sign of f_{yy}?

3. For this problem, look at the toothpicks already inserted in the plastic covering of the football.
   (a) What happens to the slopes of the toothpicks, f_x, as you move in the positive y direction along the line drawn on the football? Does the slope increase or decrease? What does this tell you about f_{xy}?
   (b) How does your observation of f_{xy} compare to your observations of f_{xx} and f_{yy}?
   (c) If the model were a cylinder instead of a football, what would be the sign of f_{xy}? How can you tell?

4.7 Partials, Gradients, and Lagrange Multipliers on a Pringles® Chip

Julie Barnes, Western Carolina University

Concepts Taught: graphing three dimensional functions, partial derivatives, gradient vectors, critical points, extrema, Lagrange multipliers

Activity Overview

In multivariable calculus, students learn a number of techniques for using partial derivatives to classify extrema and for finding extrema along a constraint curve. In this activity, students compute partial derivatives, gradients, critical points, and local extrema, as well as use Lagrange multipliers. They sketch a two-dimensional contour plot related to the surface including vectors and paths associated with their computations, and they use decorator's icing to draw the vectors and paths on a Pringles chip corresponding directly to sketches on the contour plot. Then, they analyze the meaning of these concepts on the two dimensional contour plot while making connections by exploring the same concepts on the chip.

Supplies Needed (per group)	Class Time Required	Group Size
3 Pringles chips* 1 tube decorator's icing	45 minutes	2-4 students

*Note that Pringles are used because of the unique shape that is consistent throughout all chips.

Running the Activity

Provide each group with three Pringles chips, a tube of icing, and a handout. Ask them to work through the handout. As students work back and forth between sketching various items on a contour diagram and using icing to mark those quantities on the chip, walk around the room to provide assistance.

The first problem asks students to sketch a contour diagram for the surface of the chip as seen in Figure 4.10a. Some students find this difficult and may need assistance. If they have access to three dimensional graphing software, they may compute level curves using the equation provided and compare this to their own drawings. Once their diagrams are complete, they are asked to look at a particular point and observe how the signs of partial derivatives are related to the rise and fall in elevation as well as the concavity of the surface in the given directions. They use icing on the surface, as seen in Figure 4.10b, to help them compare the vector on the surface to the corresponding vector on the contour

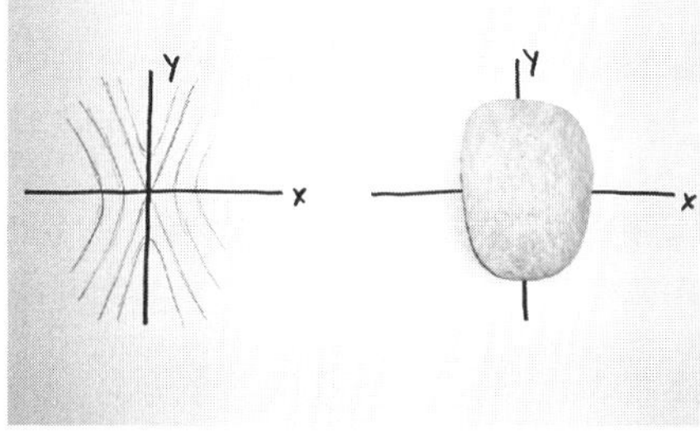

(a) Contour lines for a chip surface.

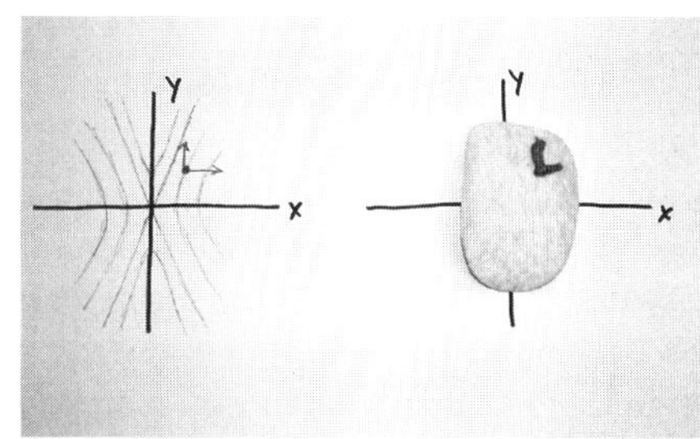

(b) Vectors tangent to the surface in the $\vec{i}$ and $\vec{j}$ direction based at $(1, 1)$.

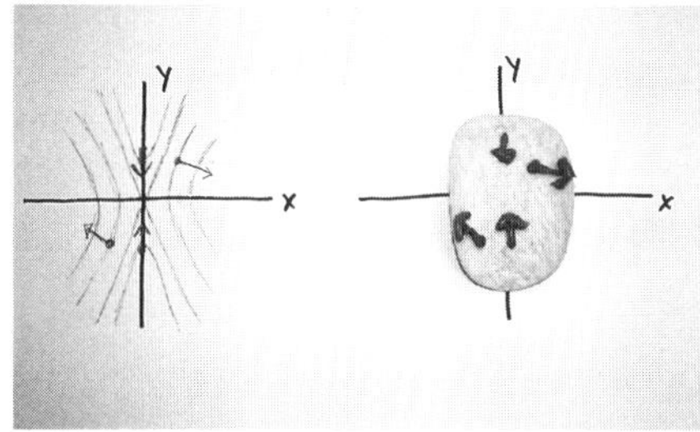

(c) Gradient vectors in the domain space, and icing vectors tangent to the surface in the gradient direction.

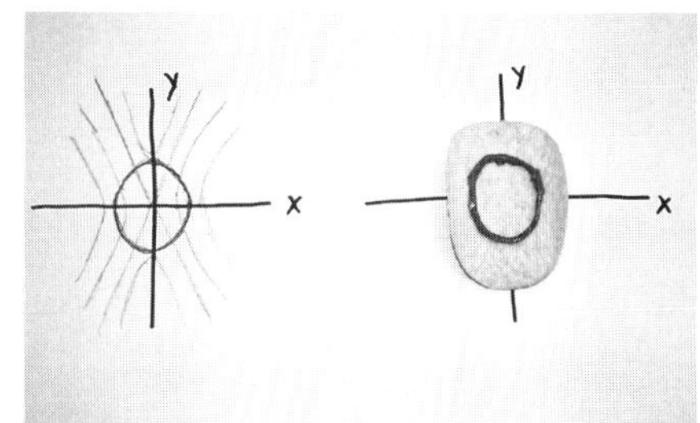

(d) Constraint equation $g(x, y) = x^2 + y^2 = 1$ in the domain, and an icing curve on the surface above $g(x, y) = 1$.

Figure 4.10: Images showing partial derivaties, gradients, and a constraint curve described in the handout.

plot. The second problem has students use analytical methods to classify the only critical point (saddle point) and identify that the physical shape is also a saddle. The third problem compares the notion of two dimensional gradient vectors in the domain space to the related vectors along the surface. The goal of this problem is for students to notice that gradient vectors are perpendicular to the contour lines and that the related vectors on the surface, as shown in Figure 4.10c, move in the direction of steepest ascent. In the final problem, related to Lagrange multipliers, students will draw a curve on their chip similar to the one seen in Figure 4.10d and will hopefully notice that the points along the curve where $f(x, y)$ reaches a maximum or minimum occur at precisely the points along the constraint $g(x, y)$ where the gradient vectors for $f(x, y)$ and $g(x, y)$ are parallel. Depending on the background of your students, you may want to pause and formalize Lagrange multipliers before students begin Problem 4d.

One of the benefits of this activity is that students can physically hold the surface they are studying. With computer algebra systems, it is possible to spin a graph and look at different angles, but nothing compares to being able to hold the surface and spin it yourself. Encourage the students to pick up the chip and look at it from different angles. Ask them to imagine where there might actually be a horizontal tangent plane or what it would be like to physically walk along the icing they drew on the chips.

Suggestions and Pitfalls

The handout is set up as an exploratory activity covering many topics, but it could also be used as a review activity or broken up into three pieces and used at different times throughout the course. It would be natural to break it into 15 minutes for Problems 1 and 2 covering partial derivatives and local extrema; 10 minutes for Problem 3 covering gradients; and 20 minutes for Problem 4 covering Lagrange multipliers.

Since icing can get a bit messy and chips can become greasy, it may be helpful to have hand wipes available in class.

Partials, Gradients, and Lagrange Multipliers on a Pringles® Chip – Class Handout

Your group needs at least three chips, a tube of decorator's icing, paper, and a pencil.

1. Draw two xy-coordinate systems side by side. On one, place the Pringles chip so that it can be reasonably modeled by $f(x, y) = 2x^2 - y^2$. On the other coordinate system, sketch the contour lines for the surface of the chip.

   (a) Compute f_x, f_y, f_{xx}, f_{yy}, and f_{xy} at the point $(1, 1)$.

   (b) At $(1, 1)$ in the domain space, sketch vectors in the $\vec{i}$ and $\vec{j}$ direction.

   (c) Use icing to represent the vectors tangent to the surface and above the $\vec{i}$ and $\vec{j}$ vectors based at $(1, 1)$.

   (d) Determine if $f(x, y)$ is increasing or decreasing as you move in the $\vec{i}$ and $\vec{j}$ directions as well as the concavity of the surface there. Compare these to the signs of $f_x(1, 1)$, $f_y(1, 1)$, $f_{xx}(1, 1)$, and $f_{yy}(1, 1)$. What can you conclude?

2. Use the partial derivatives to find all critical points analytically, i.e., places where there is a horizontal tangent plane. Describe what the surface looks like at all critical point(s). Is there a local maximum, minimum, or saddle there?

3. Draw two new xy-coordinate systems. On one, place a new Pringles chip so that it can be reasonably modeled by $f(x, y) = 2x^2 - y^2$ as before. On the other, re-sketch the contour lines.

   (a) Compute the gradient vectors at $(1, 1)$, $(0, 1)$, $(-1, -1)$, and $(0, -1)$.

   (b) At each of these points, sketch the gradient vector on the contour diagram.

   (c) Use icing to represent the vectors tangent to the surface and above each of these gradient vectors.

   (d) Describe how the gradient vectors are situated with respect to the contour lines. Compare this to what is happening to the vectors along the surface. What conclusions can you make?

4. Draw two new xy-coordinate systems. On one, place a new Pringles chip so that it can be reasonably modeled by $f(x, y) = 2x^2 - y^2$ as before. On the other, re-sketch the contour lines.

   (a) Draw the curve $g(x, y) = x^2 + y^2 = 1$ on the contour diagram.

   (b) Use icing to represent the curve on the surface and above $g(x, y) = 1$. Locate the places along the icing curve where the surface has the highest elevation and lowest elevation, i.e., the maximum and minimum values of $f(x, y)$. Mark the corresponding points on the contour diagram.

   (c) At the points you just marked, sketch the gradient vectors for $f(x, y)$ and $g(x, y)$. Describe what you see. Does this property occur anywhere else along the constraint curve $g(x, y) = 1$?

   (d) Use the method of Lagrange multipliers to find where $f(x, y)$ reaches a maximum and minimum along the curve $g(x, y) = 1$. Compare this to your observations in the previous problems.

4.8 Volume Estimation Using a Sheet Surface

Jessica Libertini, Virginia Military Institute

Concepts Taught: Riemann sum, volume estimation, integration

Activity Overview

As we begin to explore multivariable integration in the form of iterated integrals, some students have difficulty visualizing what is meant by the volume "below" a surface, or the volume between a surface and the xy-plane. In this activity, students stand on a grid with a sheet over their heads, creating a unique surface. Using their heights, students approximate the volume between the sheet and the floor.

Supplies Needed	Class Time Required	Group Size
Sheet* Tape Measure** (optional) Digital camera	15-20 minutes	Class demonstration with 12 volunteers

*A large unfitted sheet is best, but alternatively, you can use a large rectangular tablecloth. You may need to resize the number of students in your group depending on the size of your "surface."
**You may also want to have a tape measure handy in case a student does not know his/her height.

For 1 ft $\times$ 1 ft floor tiles, have the students arrange themselves with two foot spacing, such that there is sufficient space between them in each direction. If the floor is not tiled, you may want to prepare a grid on the floor with tape, such that each student has a clearly defined 2 ft $\times$ 2 ft region in which to stand.

Running the Activity

Provide students with a handout that guides the class through the activity. Have a group of 12 volunteers stand on a 2 ft $\times$ 2 ft space in a 3×4 grid. With the help of additional volunteers, drape a sheet over the heads of the students, such that your class generates a surface similar to the ones shown in Figure 4.11. Have the rest of the class look at the surface, and take some digital pictures that can be shared with the whole class, so that everyone, including the participants under the sheet, can see the surface. Problem 1 has students record the volunteers' names and locations on the grid; it may be helpful to make a copy of this information on the board. Once this information has been recorded, your surface volunteers can take off the sheet and return to their seats. Ask each of your surface volunteers to announce their height in inches, and have the class record this on their handouts. Then have students work through the handout to estimate the volume under the sheet. When students get to Problem 4, it is worth discussing units with the students and making a class decision on whether to proceed with inches, feet, or mixed units.

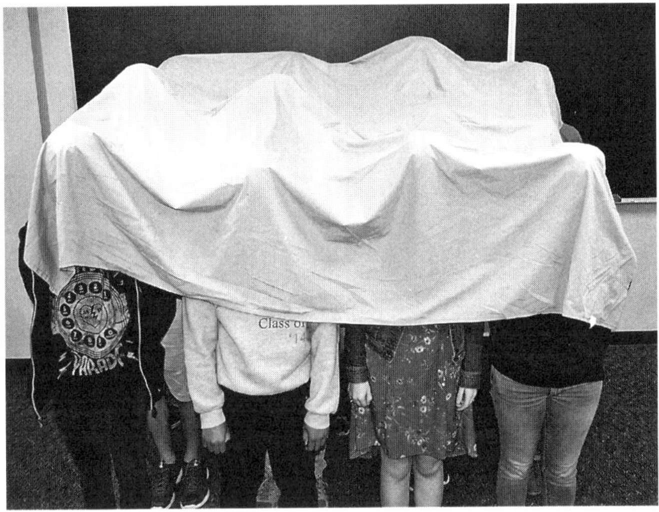

Figure 4.11: Students standing in a grid under a sheet to generate a surface.

Suggestions and Pitfalls

You may want to point out that the volume approximated in this activity is a significant overestimate as the height of the sheet quickly drops in all directions after leaving the students' heads.

Try to minimize the time students need to spend under the sheet by quickly and efficiently collecting data about who is standing where and by taking digital photos of the surface and displaying them to your class.

If you are looking to save time on this activity, take advantage of your class's collective computing power by dividing the class into groups and having each group calculate the volume for different rows or columns of students. Students can share their results with the class, and these results can then be tallied to obtain the total volume estimate. This approach also gives students a greater sense of being able to contribute directly to the collective learning experience.

For more information about this activity, as well as extensions for exploring two dimensional slices and Fubini's Theorem, see [1] where this activity first appeared.

Reference

1. J. Mikhaylov, Be the volume: A classroom activity to visualize volume estimation, *PRIMUS* 21 no. 2 (February 2011) 175–182.

Volume Estimation Using a Sheet Surface – Class Handout

1. Fill in the table below with the volunteers' names based on their locations on the grid:

2. Add the volunteers' height data to the table above.

3. We want to approximate the volume between the surface and the floor, so we will treat each person's region as a 2 ft $\times$ 2 ft square-based rectangular prism, i.e., a box. Approximate the volume of each box.

4. What units might be appropriate for calculating the volume? Why?

5. Does it matter if you first add all of the heights and then multiply by the area of each person's square base, or if you first calculate the volume of each region first and then total these volumes? Explain.

6. What is your approximation for the total volume beneath the surface?

7. Recall the appearance of the surface. Do you believe the approximation we calculated is an overestimate or an underestimate? Explain your reasoning.

8. What could you do to get a better approximation? Is there additional data you could have collected to improve the accuracy of your estimate, and if so, what?

4.9 Visualizing and Estimating the Mass of a Solid Using Multi-Colored Blocks

Monica VanDieren, Robert Morris University

Concepts Taught: visualizing a solid given a symbolic description, estimating mass, multivariable integration

Activity Overview

In multivariable calculus, some students find it difficult to determine the bounds for iterated integration. Once an iterated integral is properly set up, if a change in the order of integration is required, some students find it difficult to make adjustments to the bounds, especially if they are not visualizing the problem. Furthermore, when moving from double to triple integration, confusion concerning the integrand and its geometric interpretation arises. This activity helps students to visualize the geometric interpretation of a triple integral by drawing upon students' familiarity with Riemann sums of a function $f(x)$ over an interval I and facilitating an exploration of the three dimensional analog. Students are given a description of a solid Q in $\mathbb{R}^3$ with non-constant density. Then they construct a model to approximate the mass of the solid using colored blocks to represent different densities. Instructor-led discussion connects the symbolic expression $\iiint\limits_Q \rho(x, y, z)\, dV$ with the associated Riemann sums and the geometric model.

Using the same materials, this activity can easily be modified for double integrals to estimate the mass of a planar lamina or the volume of a solid.

Supplies Needed	Class Time Required	Group Size
Approximately 30 one$-$cm^3 color cubes in six colors*	20-30 minutes	2-3 students or Class demonstration

*If conducting a classroom demonstration, larger blocks would be more appropriate. When limited on class time, use a pre-made model constructed prior to class adhering the pieces together with a glue gun. The online store Learning Resources makes affordable plastic centimeter (and 1 inch) color cubes sold in sets of 500 or 1000. If more readily available, 1×1 LEGO® blocks may also be used; although note that the external dimensions of these blocks are 8 mm $\times$ 8 mm $\times$ 9.6 mm.

Running the Activity

Distribute the blocks and handouts amongst the groups with each group getting approximately 30 blocks in assorted colors. Next, introduce the activity to the class, emphasizing that the first task will be to construct an approximate model of the solid described on the

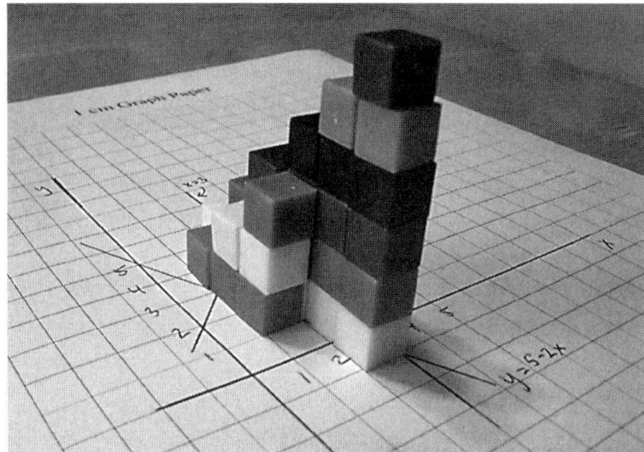

Figure 4.12: One possible model of the solid, including a layer of blocks for height $z = 0$ and using the x and y values associated with the upper right hand corner of each square on the grid.

handout. Also mention that each color will represent a different density. You may want to ask the students to visualize the problem before constructing the model by asking some questions.

> What shape do you expect the solid to be?
>
> Will the solid be more or less dense at the bottom?
>
> Will the solid be more or less dense at positions near the x-axis?
>
> What part of the solid will have the largest density?

Have the students build their models (Figure 4.12) and complete the handout. Once they have finished, lead a discussion about the activity, comparing student answers and relating the symbols to their construction. Solicit some approximations for the mass of the solid from the audience. Answers may vary since students had the option to use the upper right-hand corner, the lower left-hand corner, etc., as the x and y values associated with a square, and some students may have included a layer of blocks to represent $z = 0$. This can lead to interesting class discussion about overestimates and underestimates. Alternatively, if preferred, the handout can be reworded to disallow this ambiguity and insure that everyone gets the same model and estimate.

Next, discuss the relationship between the Riemann sum on their handout and the textbook formula for mass, $\iiint\limits_{Q} \rho(x, y, z)\ dV$. Solicit student suggestions for setting up the bounds of integration of the triple integral using the order of integration suggested on the handout. Point out that once the bounds for z have been determined, then the problem reduces to the two dimensional graph of the region R that they have sketched on their handout.

To extend the activity to a 50-minute class period, one may explore changing the order of integration. If available, use a computer projector to show a three dimensional graph of the solid with an applet such as CalcPlot3D, or sketch the solid on the board. Ask the students to use their model and the visualization of the solid to change the order of integration.

Suggestions and Pitfalls

If your class is small, you may pass around a few containers of blocks for the students to grab a few handfuls of blocks to get started. If the class size is larger, sort enough multicolored blocks into individual bags before the class to distribute to the groups.

Some students will construct a solid with only one layer of blocks; be sure to walk around the room to intervene with groups where this error is occurring. This activity will not work well in classrooms with slanted desks. Some students enjoy constructing the entire model, while others catch on rather quickly and do not need to finish the construction to make an approximation and understand the visual connection between the symbolic representation of the formula for the mass of the solid and their model.

Visualizing and Estimating the Mass of a Solid – Class Handout

For this problem we will consider the solid Q that lies under the plane $z = 2 + x - y$ and between the planes $x = 3$, $y = 2 + x$, and $y = 5 - 2x$. We will assume the units for x, y, and z are centimeters. This solid does not have constant density. Instead the density of the solid is given by $\rho(x, y, z) = x + z$ in g/cm^3.

1. Sketch a graph of the region R that lies underneath the solid Q on the 1 cm graph paper on the next page.

2. Fix a convention such as red cubes have density 1 g/cm^3 and orange cubes have density 2 g/cm^3, etc. Record your color convention below.

Density (g/cm^3)	1	2	3	4	5	6	7	8
Color								

3. Construct a model that approximates the solid Q by stacking blocks over the region R on your 1 cm graph paper. Remember to use your color convention so that your model represents not only the shape of the solid, but also the density.

4. Use your model to estimate the mass of Q. To do this, estimate the mass of each column of blocks. You can record your estimates for each column on the copy of the region R that you have sketched below. Once you have estimated the mass of each column, you can add them together to get an estimate for the mass of the entire solid.

5. In this activity, you have created an approximation for the solid Q by dividing up the solid into cubes. Use the notation Δx, Δy, and Δz for the length, width, and height of each of these cubes, respectively. For fixed values of x and y, what does the expression $\sum_{z=0}^{2+x-y} \rho(x, y, z)\ \Delta z \Delta x \Delta y$ represent?

4.10 Using a Jack to Visualize the Right Hand Rule for Cross Products

Julie Barnes, Western Carolina University
Jessica Libertini, Virginia Military Institute

Concepts Taught: cross products, right hand rule

Activity Overview

Textbooks often mention that cross products of the standard unit vectors, $\vec{\imath}$, $\vec{\jmath}$, and $\vec{k}$, can be computed using the right hand rule. However, after students read about the right hand rule and contort their right hands several different ways, some students still do not understand what the position of the right hand tells them. This activity, using a labeled jack (Figure 4.13a), helps students make connections between the right hand rule, properties of vectors, and computing cross products by taking the determinant of a 3×3 matrix.

Supplies Needed (per group)	Class Time Required	Group Size
1 labeled large plastic jack*	15 minutes	2-4 students

*Large plastic jacks are readily available online, but they can also be found in some big box stores and toy stores. Before class begins, use a permanent marker to write $\vec{\imath}$, $\vec{\jmath}$, $\vec{k}$, $-\vec{\imath}$, $-\vec{\jmath}$, and $-\vec{k}$ on the appropriate parts of each jack. See Figure 4.13a to see how the jack is labeled.

Running the Activity

In class, provide each student with a pre-labeled jack, and demonstrate how to use the jack and the right hand rule by computing a sample cross product, $\vec{\imath} \times \vec{\jmath}$. Start by holding a jack in your left hand oriented with $\vec{k}$ pointing toward the ceiling. Then place the palm of your right hand near $\vec{\imath}$ and wrap your fingers toward the $\vec{\jmath}$ direction using the most direct route possible. Call attention to the fact that your thumb ends up pointing in the same direction as the cross product, the $\vec{k}$ direction. See Figure 4.13b for a photograph of how to hold the jack and your right hand.

Break the class into groups and provide students with a handout. Have students work through the problems on the handout while you circulate through the room, assisting students as they use the jack.

The first problem in the handout asks students to compute a few cross products by taking the determinant of the corresponding 3×3 matrix. Then the second problem has

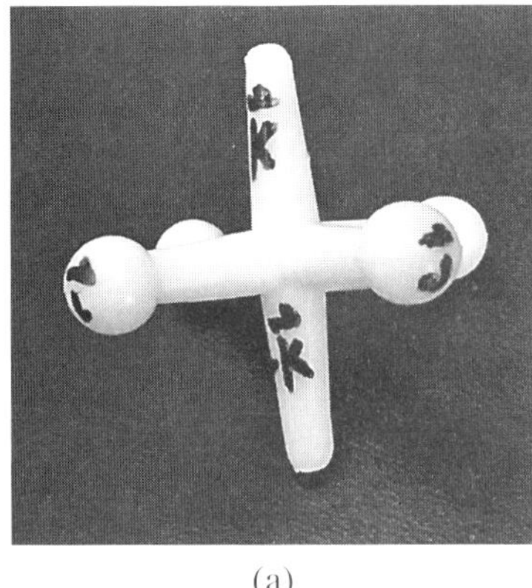

(a)

(b)

Figure 4.13: (a) A sample labeled jack. (b) A student using a jack to compute $\vec{i} \times \vec{j}$ with the right hand rule.

them compute the same cross products using the right hand rule. This should help them see that the two methods do generate the same results.

The next two problems ask students to explore the physical consequences of the right hand rule. In Problem 3, students should notice that changing the order of a cross product changes the orientation of their right hand; therefore, changing the order of the cross product simply changes the sign. Problem 4 asks students to look at how the resulting vector is positioned compared to the plane containing the original two vectors; students should notice that the resulting vector is normal to the plane.

Finally, Problem 5 asks if you need to use your right hand as opposed to your left hand. Considering that the name of the method is the right hand rule, most students will immediately say that you need to use your right hand. However, it can be valuable to have them try using their left hand to see what goes wrong.

If time allows, it is good to have a brief classroom discussion to recap the ideas covered. Many students say that going through this process helps them realize the geometric aspects of taking a cross product.

Suggestions and Pitfalls

Be sure to write on the jacks at least one day before using them in class. This will give the ink time to dry and prevent smudging.

Using a Jack to Visualize the Right Hand Rule for Cross Products – Class Handout

1. Compute the following cross products by finding the determinant of the corresponding 3×3 matrix for each: $\vec{\jmath} \times \vec{k}$, $\vec{k} \times \vec{\jmath}$, and $-\vec{\imath} \times \vec{\jmath}$.

2. Repeat the demonstration from class to make sure you understand how to use the jack to compute $\vec{\imath} \times \vec{\jmath}$. Start by holding the jack in your left hand. Then hold up the palm of your right hand near $\vec{\imath}$ and wrap your fingers toward the $\vec{\jmath}$ direction using the most direct route possible. Look at which direction your thumb is pointing to obtain the answer; in this example, your thumb should be pointing in the $\vec{k}$ direction. Now, use the right hand rule to compute the following cross products: $\vec{\jmath} \times \vec{k}$, $\vec{k} \times \vec{\jmath}$, and $-\vec{\imath} \times \vec{\jmath}$. Did you obtain the same answers as you did in the first problem? Should you?

3. When you used the right hand rule to compute $\vec{\jmath} \times \vec{k}$ and $\vec{k} \times \vec{\jmath}$, what was different about the way your hand was placed? How are the two answers related to each other? Will this always happen when you change the order in which you take a cross product?

4. Pick any two vectors represented on the jack, and denote the plane containing those two vectors by P.

   (a) Take the cross product of the two vectors. How is the cross product related to P?

   (b) Imagine taking any two vectors from P, even if they are not represented by a vector on the jack. By using the right hand rule, what can you say about how the cross product is related to P geometrically?

5. Does it matter that you are using your right hand? That is, if you held the jack in your right hand and used your left hand to compute $\vec{\imath} \times \vec{\jmath}$, do you get $\vec{k}$? Why or why not? Be sure to run a few experiments to support your answer.

4.11 Properties of Flux Using an Overhead Projector

Aaron Wangberg, Winona State University
Robyn Wangberg, Saint Mary's University of Minnesota

Concepts Taught: flux, normal vectors, divergence, divergence theorem

Activity Overview

Student comprehension of flux, which measures how much a vector field flows through a surface, relies upon understanding how the vector field interacts with the surface normal. When taught abstractly, flux can be a difficult concept. This activity illustrates an application of flux and uses a projector and a thin, flexible sheet of plastic to let students discover the relationships between flux and surface normal vectors. Challenging scenarios encourage students to propose and test conjectures which may ultimately lead to the divergence theorem.

Supplies Needed	Class Time Required	Group Size
Prepared sheet of paper* Large sheet of plastic** Projector and document camera	25 minutes	Class demonstration

*Write two rows of symbols on the paper in black marker (see Figure 4.14a).
**Use a large (16 × 24 inch) flexible (0.03 inch or smaller) sheet of plastic; if the plastic comes with a protective film, you may need to leave it attached. The symbols, when projected onto the wall through the plastic, should disappear as the plastic is angled more than 45 degrees to the light's path. The plastic we use is available from www.interstateplastics.com/Petg-Sheet-PTGCE.php, and we leave one layer of protective film attached.

Running the Activity

This activity is conducted as a whole-class exploration. Pass out the handout prior to the start of the experiment, so students can record their predictions, observations, discoveries, and notes. To begin, dim the classroom lights and project the symbols using the document camera and overhead projector (Figures 4.14a and 4.14b). While keeping the center of the film about two feet from the projector, demonstrate how the plastic sheet can be tilted or rotated between the projector and the screen, as shown in Figures 4.14b and 4.14c. The students can now discuss their observations and record their responses to Problem 1 on the handout.

Problem 2 challenges students to demonstrate the following tasks; let them take turns testing their ideas.

- If we treat the plastic as a rigid plane, can we obscure the middle column of symbols but not the others? (This is not possible: Only rotations or translations of the plastic are allowed, and these motions impact the visibility of all symbols uniformly.)
- If we treat the plastic as a flexible sheet, which can bend, can we obscure the middle column of symbols but not the others? (This is possible. The plastic in Figure 4.14d is bent into an "S" so that the plastic is parallel to the light for the middle column but perpendicular for the edge columns.)

(a) (b) (c) (d)

Figure 4.14: (a) A document camera (b) projects symbols through plastic to the screen. (c) Rotating or tilting the plastic affects visibility (flux) uniformly; (d) curving the plastic varies the visibility.

Once students figure out how to obscure the middle column, ask if anyone can explain why some symbols were more obscured. Students suggest many theories. When students refer to the plastic's position in relation to the projector's light beam, ask them to rephrase their observation using the normal vector to the surface ($\hat{n}$ or $d\vec{A}$) and the light's vector field ($\vec{V}$). Flux (i.e., the visibility of the symbols, $\vec{V} \cdot \hat{n}$) is greater when the surface's normal vector $\hat{n}$ is parallel to the vector field $\vec{V}$.

Some students need additional proof of this relationship, so in Problem 3, students experiment with the plastic to obscure the bottom row of symbols. Give students time to individually answer Problems 4 and 5 before allowing them to experiment and discover that bending the plastic into a "J" and a "C" accomplishes these tasks.

Problems 6 and 7 on the handout further explore how flux changes related to the surface normal. Ask students to make a prediction about the visibility or fuzziness of the symbols if the plastic is moved further from the projector. Have them share their reasoning prior to experimenting. Using only their eyes, have students qualitatively measure the change in the symbol's visibility on the classroom wall while a volunteer conducts the experiment. The symbols' visibility should remain constant.

Lastly, Problem 8 initiates a more formalized class discussion on the major ideas of the divergence theorem.

Suggestions and Pitfalls

Experiment with the classroom lighting, the size and darkness of the symbols, and the number of protective sheets on the plastic before class. Warning: once removed, the protective sheets are very difficult to replace.

For Problems 6 and 7, students may suggest dusty air or imperfections in the plastic impact the visibility of the symbols on the board. This is a valid issue. However, these imperfections are probably uniformly distributed, eliminating this as a cause of variations in flux.

For Problem 8, in the last experiment, since all of the light entering the plastic hits the wall, the flux per volume remains constant. Light vector fields have zero divergence away from their sources: the vector field expands radially but its magnitude decreases proportionately. To help students understand how flux informs us of changes in the vector field, draw similar experiments on the whiteboard using vector fields with positive, negative, or zero divergence. Help students discover connections between the key concepts of the divergence theorem (relationship between flux (through all boundaries), enclosed volume, and divergence of a vector field) by asking if flux will be greater on the plastic sheet or on the board in each experiment. This activity leads directly into a discussion of the divergence theorem.

Properties of Flux and Divergence – Class Handout

1. Describe the visual effect on the projection of symbols on the wall when the plastic sheet is rotated.

2. Describe how to position the plastic sheet to obscure the projection of the middle column of symbols.

3. Describe how to position the plastic sheet to obscure the projection of the bottom row of symbols.

4. Describe how to position the plastic sheet to obscure only the first column of symbols while leaving the second and third columns unobscured.

5. Describe how to position the plastic sheet to obscure the first and third columns of symbols while leaving the second column unobscured.

6. Predict whether the symbols will appear more visible, less visible, or remain the same when projected on the wall as the plastic sheet is moved farther from the projector. Explain your prediction.

7. Was your previous prediction accurate? Explain what you think is happening to the projection as the plastic sheet moves farther from the projector.

8. Discuss what results from our last experiment tell us about how the vector field changes. Explain how your experimental results today relate to the ideas of flux and divergence.

4.12 Vector Analysis of a Pop-Up Page

Vidhya Kamdar, Parsons School of Design, The New School
Jennifer Wilson, Eugene Lang College of Liberal Arts, The New School

Concepts Taught: coordinate systems, vector analysis, spatial reasoning, modeling

Activity Overview

In this activity, students deepen their understanding of three dimensional coordinate systems and apply the use of vectors to the construction and analysis of a pop-up structure as found in children's books. The activity guides students in the creation and exploration of a simple pop-up page that they investigate using basic geometric concepts, the distance formula, the law of cosines, and the algebra required to solve a system of simultaneous, non-linear equations. Students test the results of their analysis by comparing their answers to the physical model and then determine the dimensions for the *net*, a geometric solid that fits into the negative space created by the pop-up (Figure 4.15). They then construct the net and fit it into the pop-up page.

Supplies Needed (per group)	Class Time Required	Group Size
1 sample *pop-up page* and *geometric net**	1 hour (breakdown below)	2-3 students
1 piece of $8\frac{1}{2}$" $\times$ 11" piece cardstock	Part I (Problems 1-5) 15 minutes	
1 ruler per group	Part II (Problems 6-9) 25 minutes	
1 protractor per group	Making the net (Problems 10-12) 20 minutes	
1 pair of scissors per group		
Copy of full-scale net on cardstock per group		
A few rolls of tape		

*Prior to class, the instructor should complete the activity to create a sample pop-up page and geometric net. A full-scale template for the net is provided following the student handout.

Running the Activity

At the start of class, show the students your sample pop-up and ask them to think about the solid that fills the negative space; do not yet show them the net. Pass out the handout which is divided into two parts. In Part I, students construct the pop-up page and discuss the geometry of the shape created in the negative space as the page is opened and closed. See

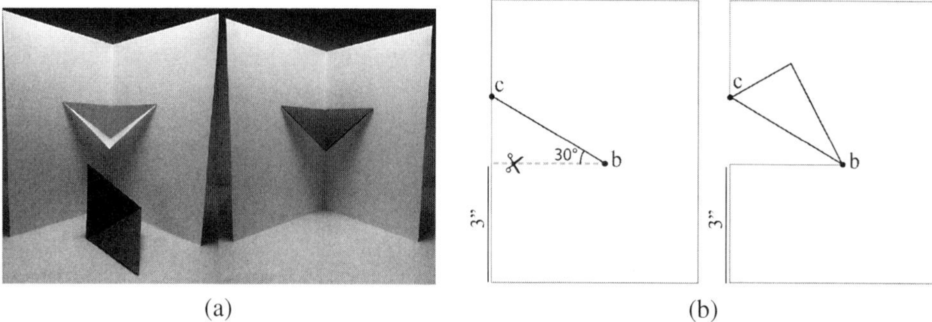

(a) (b)

Figure 4.15: (a) Pop-up page with and without geometric solid. (b) Instructions to create pop-up page.

Figure 4.15 for an example. In Part II, the pop-up page is opened to a fixed angle $\theta = 90°$, and students use vector analysis to find the dimensions of the net. The net has six triangular faces, each with vertices in the set O, P_1, P_2, P_3, and P_4 (see Figure 2 on the handout).

Part I Pop-Up Page Construction. Show the class the sample pop-up page and discuss briefly how it creates a negative space as it is opened and closed. Introduce the activity and hand out the cardstock, rulers, scissors, and protractors to the groups. Have them complete Part I of the handout to create their own pop-up page, and allow the groups time to discuss the problems in Part I. Then summarize as a class.

Part II Geometric Solid Construction. Before starting Part II, discuss the importance of defining a coordinate system, and introduce the one shown in Figure 2 of the handout. Point out that the origin is at the bottom of the protrusion and that the right-hand side of the pop-up page lies in the yz-plane with the fold along the z-axis. The left-hand side is opened at an angle θ; when $\theta = 90°$, it lies in the xz-plane. Have the groups complete Part II of the handout. Problems 6 and 7 ask for the coordinates of the vertices of the net. Points $P_1 = (0, 3, 0)$, $P_2 = (3, 0, 0)$, and $P_3 = (0, 0, 3\tan 30°)$ are straightforward to determine. The coordinates of P_4 can be determined using the fact that $\|\overrightarrow{P_4P_1}\| = \|\overrightarrow{P_4P_2}\| = 3$ and $\|\overrightarrow{P_4P_3}\| = 3\tan 30°$. Have students set up a system of three equations to solve for the coordinates $P_4 = (x, y, z)$.

The three constraints yield equations which, when squared, are stated below.

$$\begin{aligned} x^2 + (y-3)^2 + z^2 &= 9 & \text{or} \quad & x^2 + y^2 + z^2 = 6y \\ (x-3)^2 + y^2 + z^2 &= 9 & \text{or} \quad & x^2 + y^2 + z^2 = 6x \\ x^2 + y^2 + (z - 3\tan(30°))^2 &= 9\tan^2(30°) & \text{or} \quad & x^2 + y^2 + z^2 = 6z\tan 30°. \end{aligned}$$

Equating the expressions for $x^2 + y^2 + z^2$ gives $x = y = z\tan 30°$. Substituting these relationships into the first equation yields $x = y = \frac{6}{2+\cot^2 30°}$ and $z = \frac{6\cot 30°}{2+\cot^2 30°}$, or $(x, y, z) = \frac{6}{5}(1, 1, \sqrt{3})$. (This algebra takes students time to complete. Instructors should help groups who are struggling to make sure everyone has time to complete the handout.)

In Problem 8, students find $\|\overrightarrow{OP_4}\| = 6/\sqrt{5}$ and use the law of cosines to determine the angle between $\overrightarrow{OP_1}$ and $\overrightarrow{OP_4}$ which is $\cos^{-1}[\frac{1}{\sqrt{2+\cot^2 30°}}] = \cos^{-1}[1/\sqrt{5}]$. In Problem 9, students determine the remaining dimensions of the net and transfer their results to the sample net (Figure 3 on the handout). The template is based on dividing the solid along the plane $x = y$ into two symmetric triangular prisms (see Figure 2 on the handout). All measurements follow easily from previous calculations; complete dimensions are provided on the template for the full-scale net following the handout. At this point, have the students collect the supplies they need to construct the net: a copy of the template, scissors, and a few pieces of tape. You may wish to showcase the sample net that you've constructed as they make their own. After the groups construct their nets, have them put them into their pop-up pages and work on Problems 11 and 12.

Suggestions and Pitfalls

After students complete the construction and handout questions, we suggest having the groups share their pop-ups and answers to Problem 11 and 12 with the class. It can be fruitful to have the students brainstorm to develop additional questions or problems for further exploration. These might include (i) calculating the surface area and volume of the geometric solid, (ii) investigating how these properties change with θ, or (iii) analyzing other pop-up constructions (changing the length of the slit, increasing the number of slits, changing the angles of the slits, etc.).

Pop-Up Algebra – Class Handout

I. Pop-Up Page Construction

1. Follow the instructions to create the single crease and fold pop-up.

(1) Fold the paper in half width-wise. Cut a 3" slit perpendicular to the folded edge, 3" up from the bottom of the page.

(2) Draw a line from point b to the folded edge making a 30° angle.

(3) Firmly crease over $\overline{bc}$ and fold the triangle inside of the page.

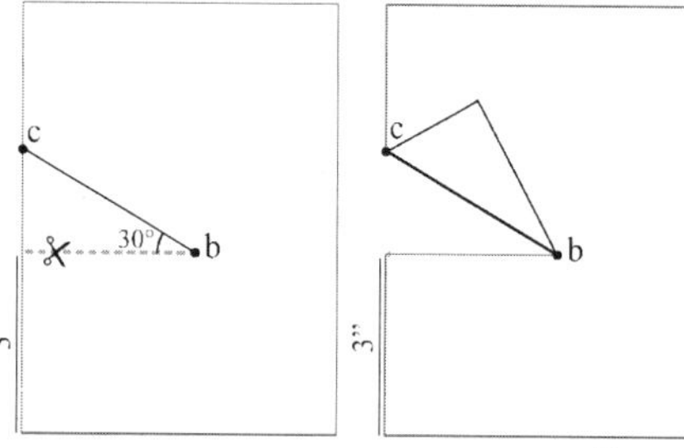

Figure 1

2. Open the pop-up and think about the solid that fills the negative space.
3. How many faces would the solid have? What shapes would they be?
4. What kinds of symmetry could you find in the solid?
5. How would the solid change as the page was opened and closed?

II. Geometric Solid Construction

6. Fill in the blanks by finding the coordinates of the vertices on the diagram.
7. Notice that the distances of $\overrightarrow{P_4P_3}$, $\overrightarrow{P_4P_2}$, and $\overrightarrow{P_4P_1}$ are fixed. Use these distances to find a system of equations and solve for (x, y, z).
8. Find the length of $\overrightarrow{OP_4}$. Use the law of cosines on ΔOP_1P_4 to find $\angle O$.
9. Fill in the lengths and angles of all of the triangles in the net below.

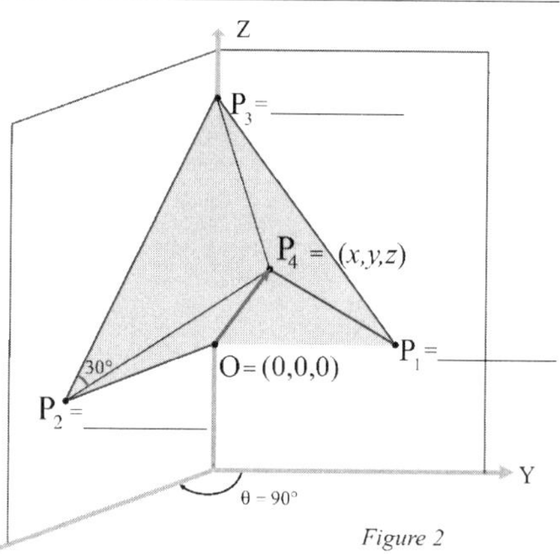

Figure 2

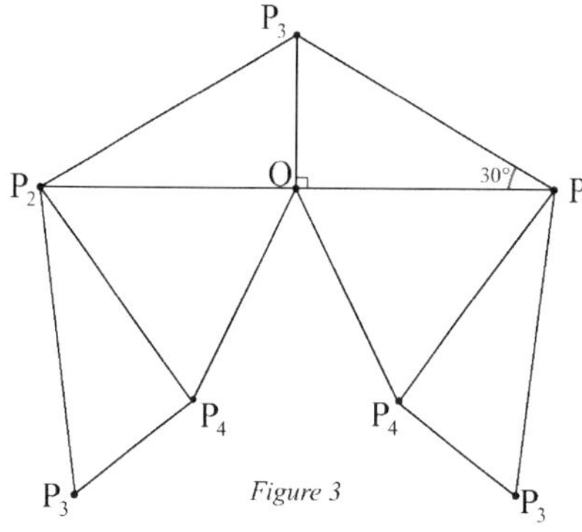

Figure 3

10. Collect the full-scale net, scissors, and tape. Construct a solid by cutting out the net and taping the tabs together. Place the solid in your pop-up.
11. How would the shape of the solid change as the page is opened? Closed?
12. How would the shape of the solid change as the angle of the crease is increased? Decreased?

Net of Geometric Solid

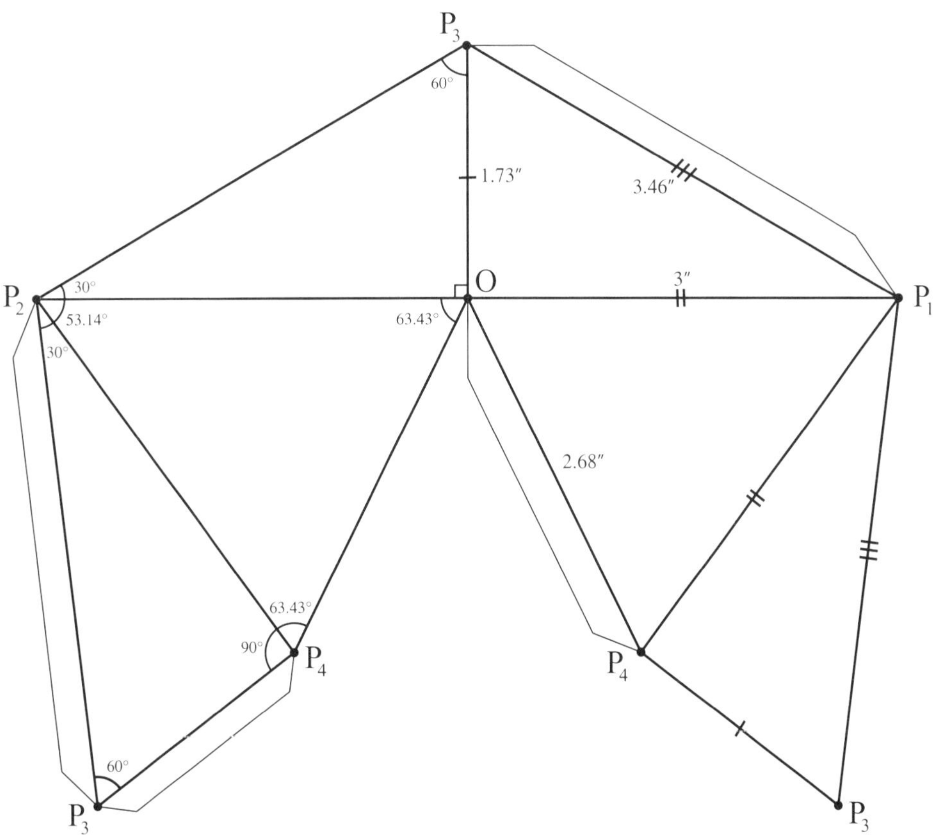

Part III

Desserts (Upper Level Courses)

Chapter 5

Sophomore / Junior Courses

5.1 Crowdsourcing to Create Slope Fields

Karen Bliss, Virginia Military Institute
Jessica Libertini, Virginia Military Institute

Concepts Taught: differential equations, slope fields

Activity Overview

A slope field is a powerful tool to visualize the family of solutions of a first-order differential equation. Typically, students use a computer to generate a slope field to avoid the repetitive and tedious task of generating one by hand. However, if students are never asked to create a slope field by hand, they may struggle with understanding its relationship to the differential equation.

In this activity, the class as a whole creates the slope field for each of three differential equations. Students each calculate the slopes for just a few points by hand, and these answers are used collectively to generate the slope field. Hence, students get an opportunity to actually compute the slope at just a handful of points, reinforcing the concept of a slope field, but through the crowdsourcing, also get to see a good visualization without the tedium and repetition required for an individual student to generate an entire slope field.

Supplies Needed	Class Time Required	Group Size
Thin strips of colored paper* Painter's tape axes** A few rolls of painter's tape or masking tape	40 minutes	Whole class

*You will need three different colors of paper, such as construction paper or card stock. For each color, cut 81 strips, each approximately $4'' \times 0.5''$.

**For each slope field, use 4 ft lengths of painter's tape each for the x-axis and y-axis. Choose a space where your class will be able to gather around to view the slope field for discussions at the end of the activity, such as a wall or a tiled floor. Write directly on the tape with a marker to indicate the grid scaling from -2 to 2 in increments of 0.5 in both x and y, each with an approximate scale of 1 ft per unit. If working on a tiled floor, you can use the floor grid to help you maintain even spacing; otherwise, you may wish to use a ruler to help maintain regular spacing. The axes can be made by the instructor while students are performing their slope calculations, or they can be prepared before class.

Running the Activity

After giving your students a brief overview of how slope fields are generated, tell them that they will be making three slope fields by hand as a class. Let them know that, since the calculation process is straightforward and repetitive, they will be working collectively to generate their slope fields, rather than each person doing all the calculations.

Each slope field will be generated on a grid from -2 to 2 with 1/2-unit spacing, meaning that there are 81 points in each grid. Assign these points to individual students in your class; this can be done using a projector or by passing out cards marked with the points. Once the students have their assigned points, provide them the handout that includes the differential equations for each of the three slope fields. Tell your students which color coordinates with each of the three slope fields, and have them record that information on the handout. Next, have your students calculate the slopes for each of the three slope fields at their assigned points and record their answers on the handout.

The handout also provides guidance on how to contribute to the construction of the slope field. In order to facilitate the construction phase, set up a central station with several rolls of painter's tape and a pile of paper strips of each color. Once students have completed their calculations, they should tape an appropriately colored strip of paper, angled to match their calculated slope, at each of their assigned points on the three coordinate systems, resulting in slope fields such as those in Figure 5.1.

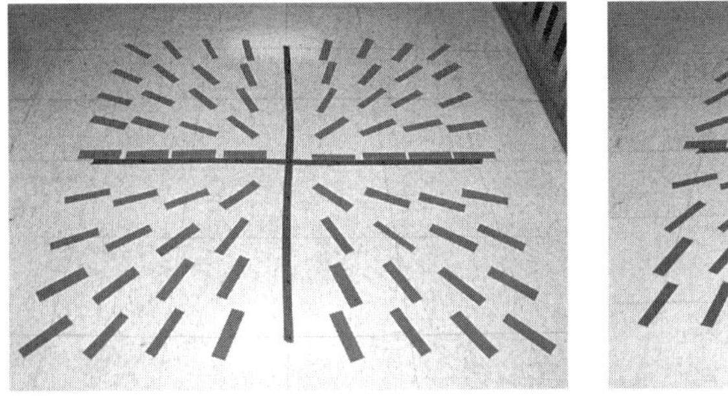

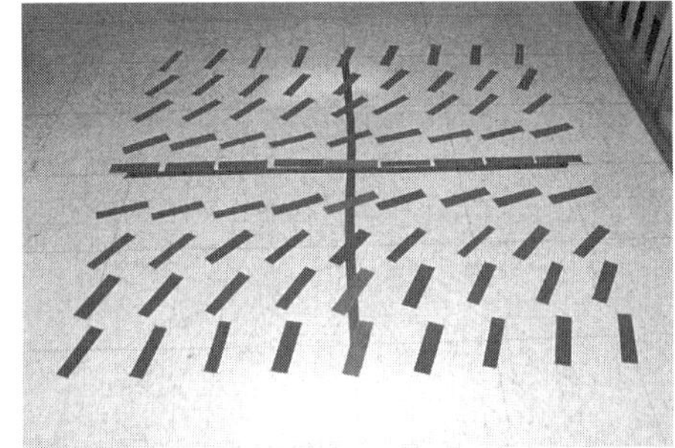

Figure 5.1: Slope fields for differential equations (A) and (C) from the handout.

After all students finish taping their paper strips onto the coordinate systems, have the students gather around the resulting slope fields. Before starting a formal discussion relating to the topics on the handout, it is beneficial to ask the class to assess and correct the slope fields; a significantly incorrect slope will stand out. Once the class is happy with each of the slope fields, lead the class in a discussion addressing the points that you would like to emphasize which could include the utility of slope fields and the types of information that can be gathered from them, families of solutions, particular solutions, equilibria, and the value of technology in generating slope fields.

Suggestions and Pitfalls

Some students work faster than others in computing the slopes, so look for ways to keep them productively contributing to the activity, such as asking these students to help with computing the slopes for the points not assigned (in case the number of points does not divide evenly into the number of students in the class), to serve as the official photographers so that the class can have a record of their work, and to facilitate the construction process as access to tape can be a limiting factor.

If time permits, you can extend the discussion portion and have students use tape or yarn to mark out particular solutions that result from different initial conditions. Also, if you wish to discuss the classification of equilibrium values, differential equation (C) offers an opportunity to see a semi-stable equilibrium value.

Clean up goes a lot faster if you ask the students to all help with removing the paper strips and tape.

Crowdsourcing to Create Slope Fields – Class Handout

You will be contributing to the class-wide generation of slope fields for each of the following three differential equations. Indicate on the line below the differential equation the color that has been chosen for that slope field.

(A) $\frac{dy}{dx} = \frac{y}{x}$ (B) $\frac{dy}{dx} = \frac{x}{y}$ (C) $\frac{dy}{dx} = y^2$

Color: ___________ Color: ___________ Color: ___________

1. Fill in the table by writing your assigned points in the leftmost column, and then finding the slope at that point for each of the given differential equations.

Assigned Point	(A) Slope of differential equation	(B) Slope of differential equation	(C) Slope of differential equation
Color of slope field:			

2. You will now use your slopes from the table to contribute to the class slope fields. On each slope field, place an appropriately colored strip of paper at each of your assigned points, tilting the strip as necessary to approximate the slope you found above. Use two pieces of tape on the back to secure each strip in place.

   Remember that a positive slope points up and to the right, while a negative slope points down and to the right. Also, a slope with an absolute value of 1 has an angle of 45°, a slope with an absolute value of less than 1 is more shallow, and a slope with an absolute value greater than 1 is more steep.

   When you have finished with your own slope strips, see what you can do to help assist your classmates with theirs.

5.2 Who's in My Differential Equations Club?

Karen Bliss, Virginia Military Institute
Jessica Libertini, Virginia Military Institute

Concepts Taught: classifying differential equations

Activity Overview

Early in a differential equations class, students can become overwhelmed by the rapid succession of all the new concepts, notation, and terminology. This activity addresses the various notations for differentiation, the identification of the dependent and independent variables, and the concepts of linearity and order for the purpose of classifying differential equations. Students form clubs where the membership is based on properties of differential equations. As students regroup themselves throughout the activity, they have the opportunity to internalize these new ideas.

Supplies Needed	Class Time Required	Group Size
Differential equation cards*	50 minutes	Whole class

*Before class, make a set of index cards or pieces of card stock each containing one of the differential equations listed in Table 5.1.

Running the Activity

To start the activity, pass out the index cards, giving one card to each student. If you have more students than cards, allow them to work in teams; if you have excess cards, keep them on hand as they will be used throughout the activity. Encourage students to think about their differential equations as you introduce notation and terminology. Start by giving a brief explanation of different notations for differentiation, including modern Newton (primes and superscripted parentheses), historical Newton (dots), and Leibniz (ratios of differentials), as all of these notations appear in the differential equations that they will be classifying. Next, introduce the idea of dependent and independent variables. It is important to stress that with some notations, especially the dot notation, that, lacking any other clues from the equation, time (or t) may often be presumed to be the independent variable.

Once the students are comfortable with these ideas, have them stand up and move around the room, forming "clubs" of differential equations having the same dependent variable. Ask them to check the cards of the other students in their group to make sure everyone belongs in that club, redirecting misplaced students as needed. If you have extra cards, have the class work together to determine the club memberships for these cards. After the club memberships have been finalized, have each club hold up their cards while a

Table 5.1: List of Differential Equations

$\frac{dy}{dx} = x + y$	$\frac{dx}{dt} = x^2t^3$	$y'' = \cos(x) + y$
$\frac{d^2x}{dt^2} + x^2 = t$	$y^{(3)} + y' + x^3 = y^4 - 5$	$y' = \sin(x)\cos(y)$
$\ddot{\theta} - \dot{\theta}\theta = 4$	$\left(\frac{d\theta}{dt}\right)^2 = \theta + 2t$	$\dot{x} = xe^t$
$x'' - x^3 = t^4$	$y'' = x + e^y$	$\frac{d^3x}{dt^3} - 3t\frac{d^2x}{dt^2} + e^t\frac{dx}{dt} + 5x = 0$
$x'' + x^{(4)} = 3x - t^3$	$y^{(4)}y' = \cos(x)$	$\frac{d^4\theta}{dy^4} = y^2 + \theta$

representative briefly tells the class his/her club's membership criterion, e.g., "We all have x as our dependent variable." If there are students who are not sure to which club they belong, have them hold up their cards and invite the class to help them find their club.

Next, have students move around the room again, forming a new set of clubs based on having the same independent variable. Repeat the process of having the students verify each others' membership and explaining the clubs' membership criteria to the class. As before, have the class work together to determine the club memberships for the extra cards.

Have the students return to their seats, and engage the class in a brief description of the order of a differential equation. Once they seem comfortable with the concept, have them get up and form a new set of clubs based on order. As before, have the students classify the extra cards, share their membership criteria, and discuss any confusion they had as they formed their clubs. Specifically, it is important here to pay attention to derivative notations, as students frequently confuse exponents with derivatives, both for Leibniz notation and superscripted parentheses.

Again, have the students return to their seats, and give them an overview of linearity. This concept can be more challenging than the other topics, so you may wish to spend more time on this explanation than on the previous explanations. Once you believe they are confident in this topic, have them split into two final clubs – one for linear differential equations and one for nonlinear differential equations. As in the previous rounds, have the students confirm one another's membership in the club and classify the extra cards. If there are students who are unsure of the linearity of their differential equations, then have the class help them.

After completing this last round of club-forming, pass out the handout, which asks students to summarize the definitions for the day's terminology in their own words, then to build their own differential equations based on their new vocabulary.

Suggestions and Pitfalls

We suggest 50 minutes for this activity because we pause multiple times to introduce new concepts. The activity could be modified to be used as a review of these concepts, perhaps before an exam, in which case it would not require as much class time.

If you plan to cover separation of variables relatively soon after this lesson, you may also want to keep the first order cards handy and repeat this activity for separability.

Who's in My Differential Equations Club? – Class Handout

Answer the following based on what you learned in today's activity.

1. Explain each of the following terms in your own words.
   i. Dependent variable
   ii. Independent variable
   iii. Order
   iv. Linearity
2. For each of the following, write a differential equation that satisfies the given criteria.
   i. x is the dependent variable, t is the independent variable, third order, nonlinear
   ii. y is the dependent variable, t is the independent variable, second order, linear
   iii. y is the dependent variable, x is the independent variable, second order, nonlinear
   iv. v is the dependent variable, w is the independent variable, first order, linear

5.3 Population Modeling Using M&M's®

Dina Yagodich, Frederick Community College

Concepts Taught: differential equations, modeling, exponential decay

Activity Overview

Many traditional differential equations courses start by introducing terminology and techniques, and therefore students do not develop an appreciation for the value of the field until much later in the course when they are exposed to applied problems. For instructors who want to motivate the subject earlier in the course, this modeling activity is designed to provide that motivation and can be done on the first day or early in the course without any prerequisite knowledge. In this activity, students will explore population changes for several scenarios by making predictions, running experiments using M&M's to simulate death and immigration, collecting data, and developing discrete models to describe the population. Students are guided through the creation of a differential equation which is then solved using MATLAB or a similar software package. The solution is graphed and compared to the data the students collected. A version of this activity that does not require the use of technology is given in the Suggestions section on page 174.

Supplies Needed (per group)	Class Time Required	Group Size
50 usable* regular M&M's 2 small cups**	10-25 minutes	2-3 students

*For this experiment to work well, each candy must be fairly symmetric and have a clearly visible "m". Before starting the experiment, students should remove any deformed candies, candies missing an "m", yellow candies as the white 'm' can be hard to see, and any extra candies so that they have exactly 50. Although buying a large bag is often more cost effective, you can save a bit on setup time by giving each group one individual (1.69 oz) size bag of M&M's as each bag typically contains 55-60 candies. Note that peanut and pretzel M&M's do not work as they are too round to definitively land "m"-up or "m"-down. Also Mini-M&M's bounce and scatter more than regular sized M&M's and may be more difficult to control; if you want to use these, you should experiment on your own before having your students do the activity.

**Each group has two cups, one with an 'X' on it for the "deceased" M&M's.

Running the Activity

At the start of the activity, split the class into groups of two or three. Give each group their M&M's and cups along with copies of the handout. Have students work through Part I of the handout while you circulate around the room. You may need to remind your students to complete their predictions before they start the experiment.

When the groups are finished, discuss what type of curve their data seems to model (exponentially decreasing) and what the equilibrium value is (zero). Explain that this curve is the solution (the result) of a differential equation. Lead the class in a discussion in which they develop the differential equation. The class may feel more comfortable starting with a discrete equation where $1/2$ of the population is lost in each generation. This leads to the discrete equation $\frac{\Delta y}{\Delta t} = -0.5y$, and since $\frac{\Delta y}{\Delta t}$ approaches $\frac{dy}{dt}$, the resulting differential equation becomes $\frac{dy}{dt} = -0.5y$ with the initial condition $y(0) = 50$.

Take this opportunity to discuss the classification of this differential equation (first-order linear initial value problem (IVP)). Using an appropriate software package, find the solution of this differential equation. For example, in MATLAB, the command to use is given below.

```
dsolve('Dy= -0.5*y','y(0)=50')
```

Have the students graph the solution given by the software and compare this to the data they had generated.

Once the students have drawn the conclusion that their data is well represented by the computational solution, discuss modifying the experiment to include immigration as explained in Part II of the handout. Have the class make a prediction on the equilibrium value. After the class has discussed their prediction on the equilibrium value, have students complete Part II of the handout. Then, lead a class discussion to determine the differential equation associated with the immigration scenario which is $\frac{dy}{dt} = -0.5y + 10$ with $y(0) = 50$. Verify the model by solving with appropriate software. For example, in MATLAB, the command to use is given below.

```
dsolve('Dy= -0.5*y+10','y(0)=50')
```

To finish the activity, have students take the solution calculated by the software and verify that it does, in fact, solve the model the class created.

This activity is based on activities found in [1] and [2].

Suggestions and Pitfalls

This activity can also be done without the use of technology. If students have already learned separation of variables, then they can be asked to solve the differential equations by hand. Alternatively, you can use the activity to motivate not only the study of the field of differential equations but also the learning of the separation of variables technique.

For Part I of the experiment, most students correctly deduce that the population will decay exponentially to zero, but students are less likely to correctly guess that the population will die out around the sixth toss. It is important for students to realize that there are three separate notions of the population for each case: the intuition-based prediction, the experiment, and the model. Depending on the background of your students, this can be a great place to have discussions about model data versus experimental data, the randomness of the experiment, and the fact that the model allows for partial candies in the population while the experiment only allows whole candies.

Part II often surprises students. Although the correct equilibrium population for this case is 20 candies, a common guess is that the population will level off at 10 candies. In fact, some students will want to continue their experiments in the hopes that eventually the population will level off at their predicted value. Depending on the makeup of your class, this can be a good opportunity to discuss what happens in the real world when scientists and engineers are faced with data that does not match their predictions.

To shorten this activity, you may skip Part I of the activity and go directly to Part II. However, including Part I provides an opportunity for students to build confidence as their prediction is typically correct and the model is a bit more straight forward to develop.

References

1. B. Winkel, Population modeling with M&M's, *International J. of Mathematical Education in Science* 40 no. 4 (2009) 554–558.

2. B. Winkel, 1-1-S-M&M's death and immigration student version, *SIMODE* (2014), https://www.simiode.org/resources/132. Accessed May 14, 2014.

Part I: Population Simulation without Immigration – Class Handout

Start the experiment with 50 M&M's® in a cup. This represents the population at $t = 0$, or the initial condition when time is zero. You will be tossing the M&M's multiple times to represent multiple generations. Each generation, any candy with the "m" facing up "dies" and will be discarded into the cup marked with an "X". The remaining M&M's will be tossed again, and the experiment will be repeated until you are satisfied. Before you begin the experiment, make some predictions.

1. What do you expect to occur?

2. Do you expect the population to level off? If so, at what value do you expect it to level off and how many generations do you expect it to take to reach that value?

Now begin the experiment.

(a) Toss the M&M's gently onto the table.

(b) Remove M&M's with the "m" facing up – they die. Place the dead M&M's into the cup marked with the "X".

(c) Count the remaining M&M's from that generation. Record the data, keeping track of the time (generation number).

(d) Go to step (a) and repeat. Keep track of time (generation number) and number of M&M's each time.

3. Sketch a graph of the population of M&M's versus time (generation number). What type of function does the graph resemble?

4. After the class discussion, write down the differential equation that models this population.

5. What is the solution to that differential equation and how well does its graph match your graph in Problem 3?

Part II: Population Simulation with Immigration – Class Handout

Again, start the experiment with 50 M&M's in a cup. As before, you will be tossing the M&M's multiple times and discarding any candy with the "m" facing up. However, this time (each generation), there will be 10 immigrant M&M's added. Before you begin the experiment, make some predictions.

1. What do you expect to occur?

2. Do you expect the population to level off? If so, at what value do you expect it to level off and how many generations do you expect it to take to reach that value?

Now repeat the experiment in Part I, replacing step (b) with these new instructions. Remove M&M's with the "m" facing up — they die. Place the dead M&M's into the cup marked with the "X". Take 10 M&M's immigrants from the cup marked with the "X", and add them to your population.

3. Sketch a graph of the population of M&M's versus time (generation number). What type of function does the graph resemble? What change do you see from the first experiment?

4. After the class discussion, write down the differential equation that models this population.

5. What is the solution to that differential equation and how well does its graph match your graph in Problem 3?

5.4 Modeling of Fishing and Restocking with Pennies

Elizabeth Carlson, Carroll College
Eric Sullivan, Carroll College

Concepts Taught: first order linear non-homogeneous differential and difference equations, mixing, simulation

Activity Overview

This activity motivates the study of the classic mixing problem in differential equations by having students use pennies to simulate and investigate the population dynamics of two fish species in a pond subject to fishing and restocking. After collecting their data, students develop, solve, and analyze a mathematical model that takes the form of an ordinary first-order linear non-homogeneous differential equation or difference equation. Depending on the goals of the course and access to technology, appropriate software may be used to facilitate this activity.

Supplies Needed (per group)	Class Time Required	Group Size
50 pennies 1 paper or plastic cup Access to scientific software (e.g., Excel, MATLAB, Mathematica, etc.) (optional)	20-30 min (data gathering) 20-40 min (writing and analyzing the differential equation)	2-3 students

Running the Activity

Provide each student with a handout, and split the class into groups of two or three. Give each group a cup with 50 pennies, and make sure that they have a flat surface, such as a desk or a table, to complete the simulation. Have students read the handout's introduction paragraphs thoroughly, and lead a class discussion to summarize the information in the following points to make sure that the students all understand the scenario.

The week starts with $N = 50$ fish.

The fish swim freely around the lake.

$M = 10$ fish are removed from the lake at random during the week.

$M = 10$ fish are restocked at the end of the week. $M/2 = 5$ of those fish are arctic grayling and $M/2 = 5$ of those fish are bull trout.

Next have the students complete the Conjecture section of the handout, and have them begin the Simulate section. The students need to decide upon a fair fishing and mixing plan that will ensure proper mixing of the newly introduced fish. After groups complete the simulation have them complete the Model section of the handout. The Analyze section of the handout can be completed either within the small groups or after groups have shared and agreed upon a proper mathematical model.

Suggestions and Pitfalls

We use coins to model the fish, but this could be done equally well with M&M's or two-sided poker chips.

We have found that this activity runs well with $N = 50$ pennies and a fishing constant of $M = 10$. With a smaller M or a larger N the data does not rapidly approach the equilibrium solution; however, you may wish to have your students experiment with other values of M and N.

When mixing the fish some students find that it is possible to inadvertently flip one of the coins. Calling the students' attention to this fact at least encourages them to beware of this possible issue.

Students will come up with many creative ways to remove fish from the lake. The methods of fishing, restocking, and swimming are purposefully left vague in the statement of the problem to allow the students some room for creativity. A nice discussion comes from asking students why it is important to let the fish swim and how this affects the choice of fishing technique. Some groups will inevitably realize that if the mixing is thorough enough then the fishing can take place from anywhere in the lake without additional bias. Allowing the students to discuss this fact leverages their potential knowledge of randomization and random sampling from statistics.

If students are having trouble with the mathematical model, then we find that directing them to a partial model such as $\Delta B/\Delta t \approx B_{new} - B_{old} = (B \text{ removed via fishing}) + (B \text{ added via restocking})$. In particular, the difference and differential equations which naturally occur from this simulation are:

$$B_{n+1} - B_n = -\alpha B_n + \frac{\alpha}{2} \quad \text{and} \quad \frac{dB}{dt} = -\alpha B + \frac{\alpha}{2},$$

where B is the proportion of species B in the lake and α is the rate at which species B is removed (for $N = 50$ and $M = 10$, $\alpha = 10/50$).

Students may initially prefer working with the total population instead of the proportion of the population. As the problem evolves it may be beneficial to discuss the advantages and disadvantages of using total population vs. population proportions. When working with proportions, the actual value of the initial population is irrelevant. Furthermore, the parameter α should approximate the growth rate of the species B population, but if we are dealing with total populations the value of α will be divided by the size of the population, making the physical meaning of the constant more difficult to see and understand. Even with these advantages, the tangible nature of working with total populations potentially makes students far more comfortable and should be encouraged until the model is nearly complete.

For more information, see [1] where this activity first appeared.

Suggested Extensions:

1. Suppose now that the Department of Fish, Wildlife, and Parks does not attempt to keep the population in the lake constant. That is, suppose that fishing reduces the population by M_1 fish each week and the Department of Fish, Wildlife, and Parks restocks M_2 fish each week. This could lead to extinction or overpopulation scenarios.
2. Consider allowing students to populate the same spreadsheet with their data (easily done with Google Sheets). This way a class-wide discussion can be had about the nature of the expected solutions.
3. Have the students plot the rate of change of population vs. the population. This should reveal a linear model and will naturally reveal the equilibrium solution as well as approximations for the terms in the differential equation.

Reference

1. E. Sullivan and E. A. Carlson (2015), 1-34-S-fish mixing student version, *SIMIODE* (2015), www.simiode.org/resources/1411.

Modeling of Fishing and Restocking with Pennies – Class Handout

A lake in northern Montana is dominated by arctic grayling (henceforth called "species A") but the Department of Fish, Wildlife, and Parks is planning to slowly introduce bull trout ("species B"). The lake is popular with sport fishermen who remove both species of fish from the lake regularly.

The Department of Fish, Wildlife, and Parks has carefully estimated the number of fish taken by sport fishing each week, and they have decided to keep the fish population as constant as possible by replacing the fish lost by an equal number of arctic grayling and bull trout. Consider a lake with $N = 50$ fish in the lake at the beginning of the week. Fishermen remove $M = 10$ fish during that week, then the fish and wildlife people will restock the lake with 5 arctic grayling and 5 bull trout. Hence the population of the lake will remain $N = 50$ fish at the end of each week, assuming no new fish are born. Both fish species swim freely throughout the lake and both are targeted by similar bait used by sport fisherman.

1. **Conjecture**

   (a) What do you think will happen to the populations of species A and B over a long period of time?

   (b) Is it possible that species A will be eliminated from the lake with the restocking plan? Explain.

2. **Simulate**

   (a) Use pennies to represent your N fish and decide with your partner(s) which coin face represents which species (e.g., Species A will be heads). Start your lake with 100% species A.

   (b) Decide with your partner(s) how to simulate the swimming of fish, the fishermen, and the Department of Fish, Wildlife, and Parks' restocking plan. Simulate roughly 15 weeks of the fish population representing species A and B with coins. Be sure to let the fish swim thoroughly around the lake and keep track of the proportions of species A and B. Keep track of your data in a table with the following headers:

Week #	Number in population		Proportion of population	
	species A	species B	species A	species B
0	50	0	1	0
⋮	⋮	⋮	⋮	⋮

3. **Model**

   (a) Propose a verbal model for the rate of change of species B in the lake.

   rate at which species B changes = ______________ + ______________

   (b) Explicitly state any assumptions that you are using in your verbal model.

   (c) Introduce mathematical notation for your proposed model and write your verbal model mathematically. Be sure to include any necessary condition(s).

   model: ____

   condition(s): ____

4. **Analyze**

   (a) According to your model, what is the long term effect on the fish population in the lake? Use your model to justify your answer algebraically and graphically.

   (b) Solve your mathematical model (either numerically or analytically) and compare with your data. How well does the model interpolate and extrapolate the data?

5.5 Modeling a Falling Column of Water

Brian Winkel, SIMIODE

Concepts Taught: mathematical modeling, data analysis, parameter estimation

Activity Overview

In mathematical modeling, students are challenged with finding relations, and frequently, it is more reasonable, though less obvious, to identify a relationship involving a rate of change. In this activity, students perform an experiment and collect time-dependent data on the height of a falling column of water. They then use their data to develop and test several possible mathematical models for this data by relating the rate of change of the height to a function of the height. Ultimately, students should recover Torricelli's Law and are given the opportunity to compare their own work with this famous result.

Supplies Needed	Class Time Required	Group Size
Apparatus for collecting data* Stopwatch and video camera** 2 liters of water Food coloring Small holed funnel	30 minutes	3 students

*You may only need one apparatus for the class; groups may take turns collecting data from it. For large classes, you may want to prepare more than one apparatus. For a single apparatus, remove the label from a clear, two-liter soda bottle. Then, use a drill or punch bit to create a small borehole near the bottom of the vertical section of the bottle. Cover the hole with masking tape. Place the bottle on a stand that is next to a collection device that is large enough to hold two liters of water as seen in Figure 5.2a. During the experiment, water flows through the hole into the container; to prevent spills it is important that the bottle is situated so that the arc of the departing water enters the collection device properly. Photocopy a metric ruler onto a piece of paper, cut it out, and tape it to the bottle so that 0 is at the same height as the borehole and numbers increase vertically. Be sure that the ruler does not cover the borehole. If all these parts are assembled ahead of time the apparatus should only take five minutes to set up in class.
**The stopwatch and video functions can also be accomplished using other devices, such as smartphones, tablets, or computers.

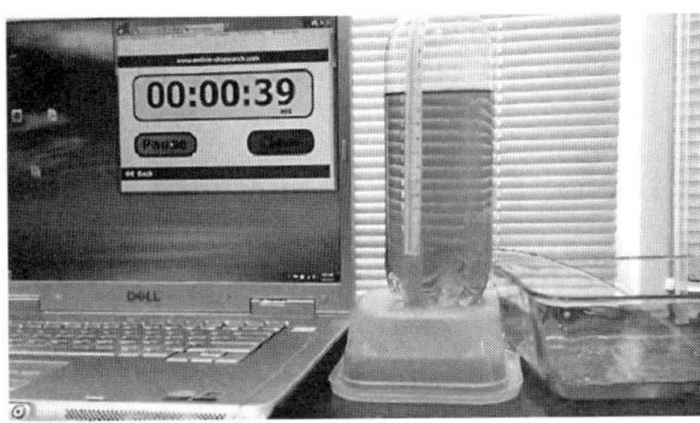

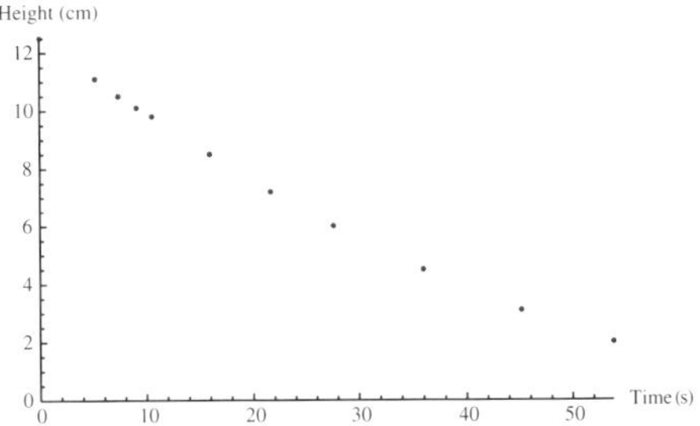

Figure 5.2: (a) An apparatus for collecting data on water flowing from a column of water. (b) Typical data plot of the height of a column of water in cm vs. time of the observation in seconds.

Running the Activity

Bring the prepared apparatus for collecting data to class and make sure you have a space for students to conduct an experiment that might spill water. Also have a container of approximately two liters of water. Add a couple of drops of food coloring to the water so that it will be easier to see.

Start by showing your class how to use the equipment to run and record their experiment. Provide each group with a copy of the handout and give them access to the equipment. Have them follow the directions on the handout to create videos of a falling column of water. Before the next group of students uses the apparatus, have have them re-tape the hole and pour the water from the collector back into the bottle using the small funnel. Be available to answer questions throughout the data collection process.

Once each video has been created, have students replay it, stop when the stopwatch lands at evenly spaced time intervals, and record the height of the water at those time intervals. After students record their data, Problem 2 has them plot the height of their column of water vs. time. The plot will look something like that shown in Figure 5.2b. This will be useful later for comparing their model to the collected data.

The most interesting relationship that can be found from this data comes from Torricelli's law. To help students make this observation, Problem 3 asks them to use their data to compute Δh, h^2, and $h^{1/2}$. Then in Problem 4, they use this data to determine whether $h'(t) = -k_1 h(t)$, $h'(t) = -k_2 h(t)^2$, or $h'(t) = -k_3 h(t)^{1/2}$ fits the data best. The easiest way for them to do this is to plot $h'(t)$ vs. $h(t)$, $h'(t)$ vs. $h^2(t)$, and $h'(t)$ vs. $h^{1/2}(t)$ to see which one is most linear. Once they determine the best line, Problem 5 asks them to find the constant, k_i, associated with that line. They now have a differential equation modeling the relationship between height and change in height of water. In Problem 6, students solve this differential equation for $h(t)$. Then in Problem 7, they plot their solution as $h(t)$ vs. time on the same axes as their original data plot from Problem 2. The final two problems ask students to research Torricelli's law and then speculate what will happen if some of the features of their testing apparatus are changed.

Suggestions and Pitfalls

Before the class, run the experiment to test that it is working properly.

If your experimental setup is not working properly, prerecorded videos of a similar experiment are available at [2]. Students can collect their data from these videos as they would have from their own videos.

For students who have modeling experience, instructors may choose to leave the activity more open-ended. For example, instead of providing students with the three equations in Problem 4, keep r as a parameter for $h'(t) = -kh(t)^r$ and have students determine the best r value. For a more open-ended experience, have students collect data, plot $h(t)$ vs. time, and leave them to determine the best kind of model to explore.

For more information, see [1] and [2] where this activity first appeared.

References

1. B. Winkel, 3-70-S-Falling in water, *SIMIODE* (2016) www.simiode.org/resources/1595. Accessed 28 January 2015.

2. B. Winkel, *SIMIODE Videos* www.youtube.com/channel/UC14lC-tyBGkDPmUnKMV3f3w. Accessed 11 September 2014.

Modeling a Falling Column of Water – Class Handout

You will be collecting data to determine if there is a relationship between the height of a falling column of water and the rate at which it falls. Be sure that you have the prepared apparatus, a stopwatch, and a device to record videos.

Be sure that the soda bottle is on top of the stand with the bore hole aimed toward the receiving container. Place a stopwatch next to the apparatus. Then place a video camera (or cell phone in video mode) in a location where both the stopwatch and ruler on the bottle are in clear view. Have your instructor check that your setup is correct. When everything is in place, start the video, start the stopwatch, and tear away the tape covering the bore hole to let the water flow. Stop recording when the water stops flowing.

1. Once you have created the video, play it back, stopping when the stopwatch reaches evenly spaced time intervals. At those times, record data on the time and height of the water in the first two columns of the following table.

Time (t)	Height $h(t)$	$\Delta h(t)$	$(h(t))^2$	$(h(t))^{1/2}$

2. Plot your data as $h(t)$ vs. time.

3. Fill in the rest of the table.

4. Use your data to find which of the following possible models best fits your data: $h'(t) = -k_1 h(t)$, $\quad h'(t) = -k_2(h(t))^2$, $h'(t) = -k_3(h(t))^{1/2}$.

5. Find the constant, k_i, for the best fit model you just found.

6. Using your best fit model and associated constant, solve the differential equation for $h(t)$.

7. Plot your solution, $h(t)$, vs. time on the same axes as the plot you created above in Problem 2. How close are these plots? Is your model accurate? Is your model reasonable?

8. Research Torricelli's law. How does this relate to this problem?

9. Consider changing the physical configuration, such as changing the cylinder diameter, borehole diameter, water column height, etc. How would this change your mathematical model? How would this change the water flow in the experiment?

5.6 Using Linear Algebra Definitions to Find Your Team

Martha Allen, Georgia College & State University

Concepts Taught: span, linear independence, basis

Activity Overview

This two-part activity is designed for an introductory linear algebra course to introduce or reinforce the concepts of span, linear independence, and basis. First, students use definitions in order to divide the class into teams. Once the students are in their teams, they undertake an in-class assignment in which they justify why they belong together as a team and then consider some broader problems on the handout to reinforce the lesson.

Supplies Needed	Class Time Required	Group Size
Team-forming cards*	30 minutes	3-4 students

*Prior to class, photocopy the team-forming cards from Table 5.2 onto colored cardstock, and cut out each line separately to obtain the deck of 16 cards.

Table 5.2: Information for Team-Forming Cards

Linearly dependent vectors in $\mathbb{R}^3$
$(1, 1, 2), (3, 5, 8), (13, 21, 34)$
$(2, 1, 3), (1, 2, 1), (1, 1, 4), (1, 5, 1)$
$(1, 2, 3), (2, 1, 3), (0, 0, 0)$

Linearly dependent vectors in $\mathbb{R}^2$
$(-3, 7), (\frac{15}{4}, -\frac{35}{4})$
$(2150, 2), (1075, 1)$
$(2, 3), (5, 7), (11, 13)$

Basis for $\mathbb{R}^3$
$(2150, 0, 0), (0, 2150, 0), (0, 0, 2150)$
$(1, 1, 1), (1, 2, 3), (0, 1, 0)$
$(2, 3, 5), (7, 11, 13), (17, 19, 23)$

Basis for $\mathbb{R}^2$
$(-3, 7), (15, 4)$
$(2150, 3030), (-1261, -1262)$
$(17, 19), (23, 31)$

Running the Activity

Have each student draw a card. Inform students that each card has either a set of vectors or a property about a set of vectors. Tell the students that their first task is to use their cards to find their team members. If a student has a card with a property, then the student is looking for other students in the class with cards that have vectors satisfying the given property. Similarly, if a student has a card with a set of vectors, then the task of that student is to find the property that his or her vectors satisfy; in addition, he/she should also find other students with vectors sharing the same property. Encourage the students to do this as quickly as possible. Once the students have found their team members, then their next task is to justify at the chalkboard why they are a team; in other words, each team explains why their sets of vectors satisfy the property of the team. The instructor provides immediate feedback on whether the team formation is correct.

After the teams are formed, each team works collaboratively on the handout in class. Problem 1 asks them to record the reasons for why they are a team. Problems 2a and 2b allow them to analyze two other sets of vectors, where at least one set is different from the group in which they belong. Problems 2c and 2d ask students to think more deeply about the topics being covered. In Problems 2e and 2f, students are asked to distinguish between a finite-dimensional vector space and the order of the set of vectors in a vector space. In other words, a vector space can be finite-dimensional but contain infinitely many vectors. In the case of 2e, the statement is true since one could give the vector space $\mathbb{R}^{12345678910111213141516171819}$ as an example and the natural basis in this vector space would satisfy the required number of linearly independent vectors. In the case of Problem 2f, the statement is false since the only vector space with a finite number of vectors is the zero vector space. This could lead to further discussion on the closure properties of a vector space, if necessary.

Suggestions and Pitfalls

The team-forming activity can be modified to accommodate a class with fewer than 16 students. If there are between 8 and 15 students, remove some of the vector cards, taking care to leave the four groupings with roughly the same number of cards per group. For example, if there are 14 students, then remove two cards with sets of vectors from different groups, which would yield two teams of three and two teams of four. If there are fewer than eight students in the class, one modification would be to divide the class into two groups, and then give each team a shuffled set of the 16 cards. The task of each team of students would be to match the cards with vectors with the appropriate property cards.

If the class size is larger than 16, then print the cards again, but this time onto a different color of cardstock. In this case, the students are not only trying to find their team members as described in the previous section, but they would also have the added condition that they are also looking for those students with cards of the same color.

Using Linear Algebra Definitions to Find Your Team – Class Handout

1. Explain why you are a team (that is, explain why the set(s) of vectors satisfy the property of the team).

2. Determine whether the following statements are true or false. Write out *true* or *false*. Justify your answers clearly.

   (a) The vectors $(1, 2), (-1, 3), (5, 2)$ are linearly independent in $\mathbb{R}^2$.

   (b) The vectors $(1, 0, 0), (0, 2, 0), (1, 2, 0)$ span $\mathbb{R}^3$.

   (c) A basis for a vector space V can include the zero vector.

   (d) There exists a set of vectors in $\mathbb{R}^3$ that is linearly independent but does not span $\mathbb{R}^3$.

   (e) There exists a vector space containing a set of 12345678910111213141516171819 linearly independent vectors.

   (f) There exists a vector space containing exactly 12345678910111213141516171819 vectors.

5.7 Picturing Prime Factorization

Mary Flagg, University of St. Thomas

Concepts Taught: factorization

Activity Overview

This activity prepares students to prove theorems on divisibility commonly found in courses such as abstract algebra, number theory, and discrete math. The fundamental theorem of arithmetic states that every positive integer greater than 1 can be written uniquely, up to the order of the factors, as a product of prime numbers. In this activity, students create a concrete visual picture of the prime numbers as building blocks for composite numbers. The students will factor a composite number n as a product of primes, illustrate this with sticky notes as shown in Figure 5.3, and then show that every factor of n is a product of some combination of prime factors of n.

Supplies Needed (per group)	Class Time Required	Group Size
A set of sticky notes in 5 shapes,* 12 notes of each shape 1 marker	15-20 minutes	3-4 students or Class demonstration

*Using notes with a different shape for each distinct prime creates a clear visual picture; shaped notes can be purchased or cut out of the standard square notes. Notes of different colors work equally well. If using colored notes, replace the reference to blocks of different shapes in the following discussion with blocks of different colors.

Running the Activity

Give each student a copy of the handout. Divide the class into groups of 3 or 4, depending on class size and room layout. Each group will need a marker, a set of sticky notes, and a space such as a desk, table, or chalkboard to arrange the sticky notes as shown in Figure 5.3a. Ask the students to make shape-coded prime factor blocks, such as those shown in Figure 5.3, by using their marker to write a large "2" on each of their square shaped notes, a large "3" on each of their triangular notes, etc., such that each of the five shapes corresponds with a distinct prime: 2, 3, 5, 7, or 11.

To illustrate the process that will be used throughout the activity, begin by working through the first problem on the handout as a class. Copy the table from the handout onto the board. Using 2-blocks and 3-blocks, fill in the table as in Figure 5.3a to demonstrate the different ways of factoring 12. Thus we see $12 = 1 \times (2 \times 2 \times 3)$, $12 = 2 \times (2 \times 3)$, and $12 = 3 \times (2 \times 2)$. Have students transcribe the numbers onto their handouts. Then have the students do the second problem on the factors of 60, shown in Figure 5.3b, and complete

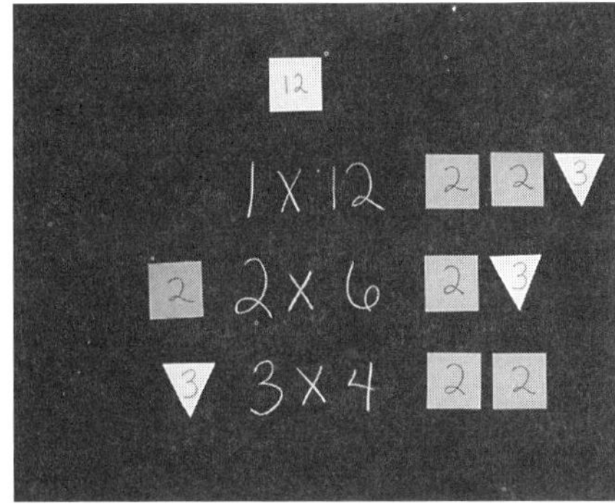

(a) Factoring 12.

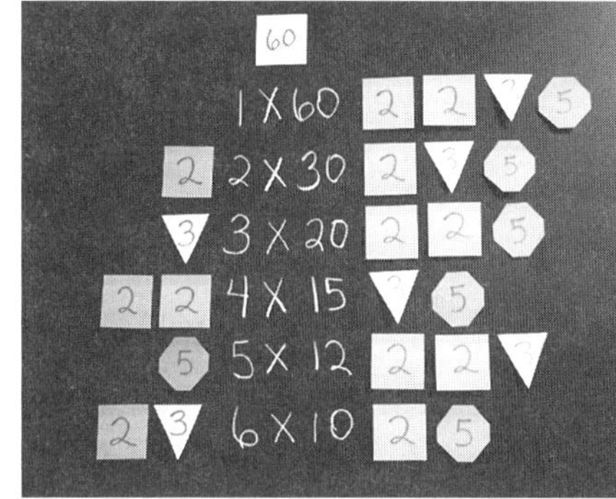

(b) Factoring 60.

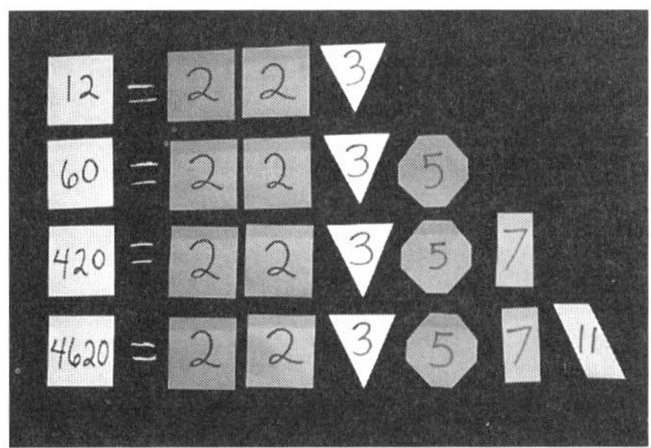

(c) Prime factorization.

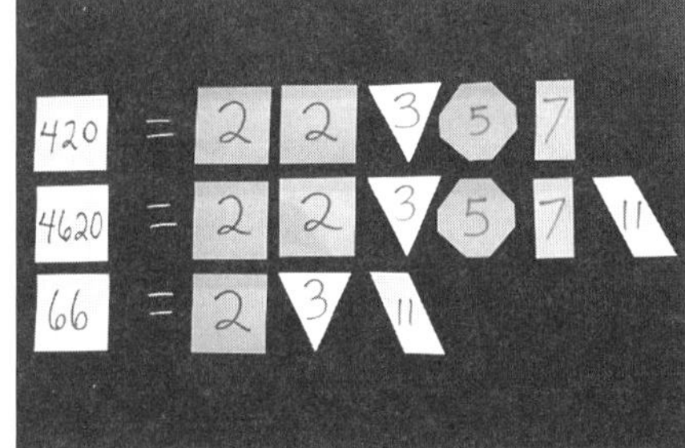

(d) Comparing factorizations.

Figure 5.3: Factorizations using sticky notes.

Problem 3 using their results. The students should notice that the factors of 60 are simply the different multiplicative combinations of the prime factors of 60.

Reinforce the idea that the factors of an integer n are multiplicative combinations of the prime factors of n by having the students complete the rest of the handout. The prime factorizations of 420 and 4620 asked for in Problem 4 are illustrated in Figure5.3c. Problem 5 asks students to compare factorizations. Figure 5.3d illustrates the answer to Problem 5a on the handout: 66 is a factor of 4620, but not a factor of 420. For Problems 5b, 5c, and 5d, students repeat the process from 5a.

Finally, Problem 6 asks students to generalize their results.

Suggestions and Pitfalls

The numbers 12, 60, 420, and 4620 work well because they are the smallest numbers with a square factor and all the smallest five primes represented. Any combination of small primes may be used. Larger primes may also be used, but it can become counterproductive to have the factoring take too long.

If the class has formally defined divisibility and the greatest common factor of two integers, the handout may be followed by a class discussion of basic divisibility properties. For example, with the factors of 12 and 60 illustrated on their work area, you could ask students to prove or disprove the following conjecture about divisibility.

Conjecture 1. For all integers a, b, and c, if $a|c$ and $b|c$, then $ab|c$.

This conjecture is false, as they should be able to see fairly easily. However, it is worth asking the students when ab is a factor of c and when it is not, using the factors in front of them. This should lead to their informal explanation of the truth of Conjecture 2.

Conjecture 2. For all integers a, b, and c with $gcd(a, b) = 1$, if $a|c$ and $b|c$, then $ab|c$.

Picturing Prime Factorization – Class Handout

Using the sticky notes as "prime blocks", illustrate the prime factorization of an integer by placing its prime factors next to each other. The operation of multiplication will be implicitly understood for blocks placed next to each other.

1. Illustrate the prime factorization of 12 and the divisors of 12 by using the prime blocks to fill in the table:

Prime Factors of a	$a \times b = 12$	Prime Factors of b
	1×12	$2 \times 2 \times 3$
	2×6	
	3×4	

2. Construct a similar table for all possible ways to factor 60 as a product of two positive integers and illustrate the prime factorization of each divisor of 60.

3. What do you notice about the relationships between the prime factorization of 60 and the prime factorizations of each of its divisors?

4. Find the prime factorizations of the numbers 420 and 4620, and illustrate their factorization using the prime blocks. Keep these factorizations on your workspace as you complete the next problem.

5. Which of the following integers is a factor of 420? Which is a factor of 4620? How do you know?

   (a) 66

   (b) 56

   (c) 70

   (d) $2 \times 3 \times 5 \times 5 \times 7$

6. For all integers a and n, if a is a divisor of n, what is the relationship between the prime factors of a and the prime factors of n?

5.8 Traffic Jam: A Lifesize Logic Puzzle

Melissa A Stoner, Salisbury University

Concepts Taught: problem solving, pattern seeking, logical reasoning, logic puzzle

Activity Overview

As students begin to develop ideas of logical thinking and the foundation of proof, they often struggle to understand the logical flow of thinking and to meld the proof structure with mathematical ideas. In this activity, we immerse students in a problem that requires no prior knowledge to solve. The problem comes in the form of a game. The correct move on the game board at any time can be determined using logical statements and can be proven. This game gives students a non-mathematical context in which to practice drafting a proof by contradiction.

Supplies Needed (per group)	Class Time Required	Group Size
9 circles*	20-40 minutes	8-10 students (two teams of 4)

*Each group will need nine circles that are approximately ten inches in diameter. Laminated card stock works best, but paper plates work great too. If desired you can use four plates of one color, four plates as another, and a neutral one in between.

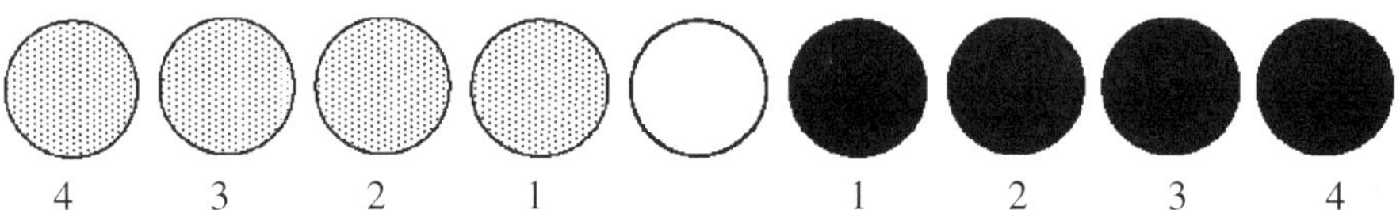

Figure 5.4: Traffic Jam set up. Note that the numbers represent the students.

The game board is set up using the nine circles. The circles are placed on the ground in a line as shown in Figure 5.4. Place the circles a foot or so apart to give students a little personal space.

Running the Activity

Begin by setting the game board to mimic the setup in Figure 5.4. Divide the class into groups of 8-10 students, each with a game board, and for each group create two subgroups of four students. Any additional team members can observe the group's movements and record information for the group; they should be active members of the discussion as they have a great vantage point to observe the movement patterns. For each group of students, have four stand on the left circles and four stand on the right circles, forming two teams. The teams will then face the opposite team. The goal of the game is for each team to move to the other team's circles. However, they must abide by these rules of the road.

1. Only one person, from either team, may move at any time. Teams do not need to alternate moves.
2. You must remain on a circle or move to an open circle. Only one person on a circle at a time.
3. You may not jump members of your own team.
4. You may only jump one person at a time.
5. You may not move backwards. If you are stuck (in a traffic jam) you must reset the game.

Figure 5.5 shows students doing the activity. Allow students time to accomplish the task; there is only one solution. Once they have accomplished the goal of switching sides, have them repeat their solution. (Some groups stumble upon the correct solution without understanding it.) Follow the activity with the class handout requiring students to answer questions, discover patterns, and explain the logical thinking in the solution strategy. At each step of the solution there are two possible moves that do not violate the rules. Only one of these is the correct choice. The correct choice can be determined by analyzing the consequences of the two options. The analysis can be proven and outlined as a proof by contradiction. To prove that choice A is correct, assume choice B and follow the subsequent moves until a traffic jam (contradiction) happens.

Suggestions and Pitfalls

Be aware that the runtime for this game often varies dramatically based on the population. Some groups who think very analytically can arrive at the solution in just five minutes, while other groups get frustrated and often require up to forty minutes to complete the game. Groups who struggle often need group management assistance or for the instructor to help them summarize what they know and how they can adjust their strategy. The activity can also be scaled up or down by varying the number of team members on each side and adjusting the corresponding number of circles.

The solution to the puzzle, using symbols A and B for movement by someone on Team A and Team B respectively, is ABBAAABBBBAAAABBBBAAABBA. Similarly, the solution is BAABBBAAAABBBBAAAABBBAAB if Team B moves first.

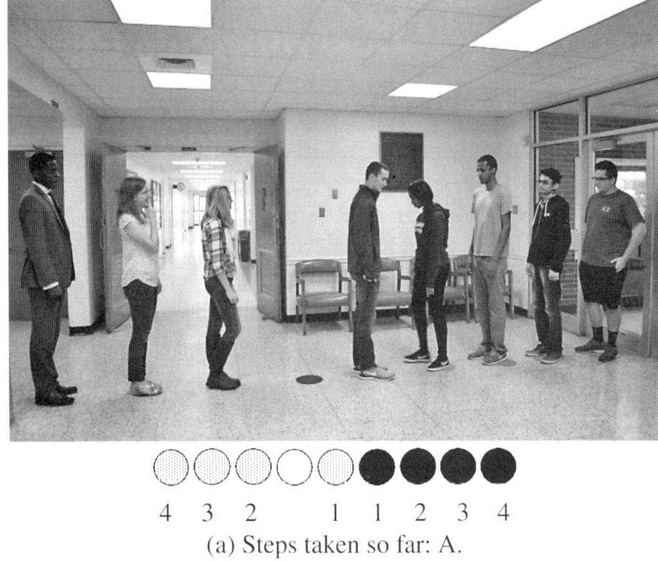

(a) Steps taken so far: A.

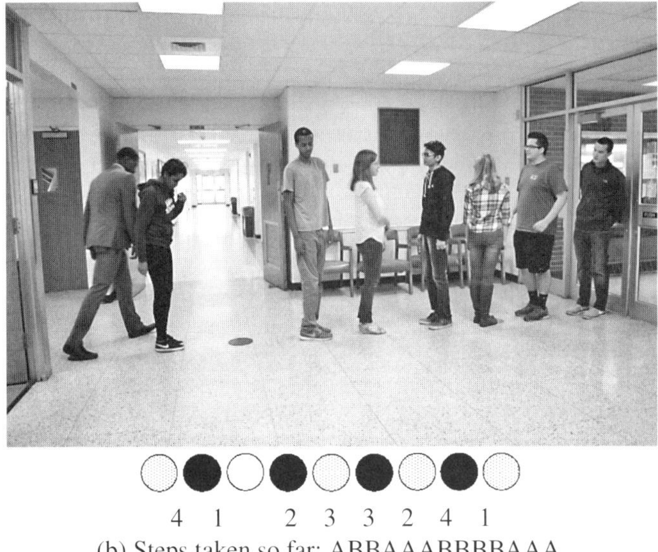

(b) Steps taken so far: ABBAAABBBBAAA.

Figure 5.5: Students solving the puzzle. The corresponding diagrams symbolically indicate where each student has moved in two different ways. (a) In this photograph, students have just begun to solve the puzzle. Only one person from Team A has moved as indicated by the 1 on the left moving one circle over, and also by "Steps taken: A." (b) Students are much further along in the puzzle and in this stage both teams are in alternating positions. To get there, someone from Team A moved, then B, then the next person from B, then A, and so forth resulting in ABBAAABBBBAAA.

Traffic Jam – Class Handout

With two teams of four facing each other on the game board, the game is completed when the teams trade places on the game board using these rules.

Only one person, from either team, may move at any time. Teams do not need to alternate moves.

You must remain on a circle or move to an open circle. Only one person on a circle at a time.

You may not jump members of your own team.

You may only jump one person at a time.

You may not move backwards. If you are stuck (in a traffic jam) you must reset the game.

1. Using sentences, diagrams, pictures, etc., summarize the solution to Traffic Jam.
2. Write a pattern of movements that can be used to represent the movements in Traffic Jam for four people on each team. (A pattern is something like ABABABBABB... where "A" represents movement from Team A and "B" represents movement from Team B.)
3. Extend your pattern of movements to one that works for five people on each team.
4. Consider the game board below. Determine the next successful move and use a proof by contradiction to justify it. *Hint: You may assume that two members of the same team residing on consecutive circles that are not their original circles will create a traffic jam.*

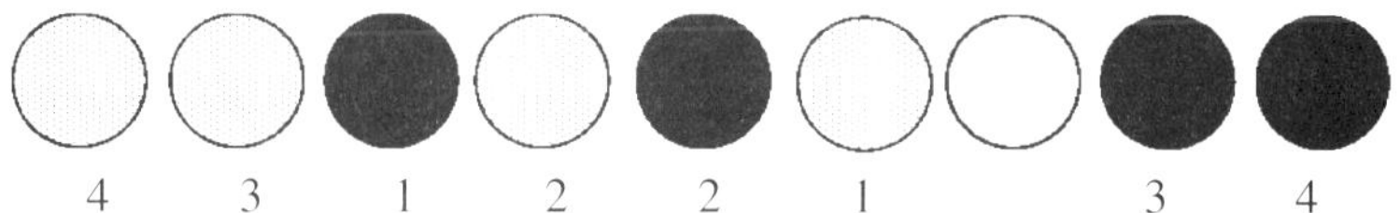

Recall that students started in this position:

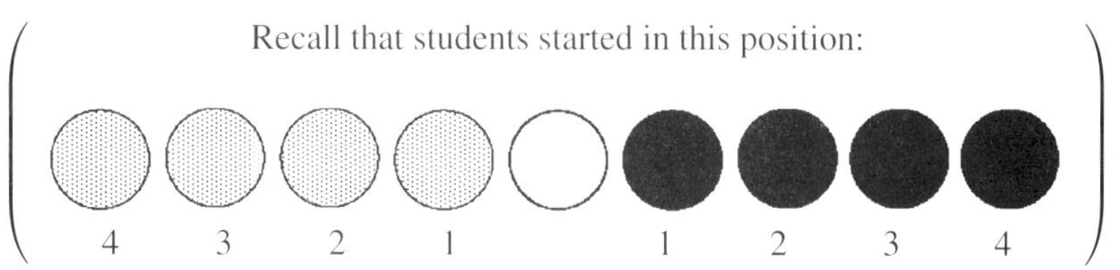

5.9 Living De Morgan's Laws

Mary Flagg, University of St. Thomas

Concepts Taught: De Morgan's laws, set operations

Activity Overview

De Morgan's laws, which describe the complement of a set union or a set intersection in terms of the complements of the individual sets, are a standard part of basic set theory and propositional logic topics in a wide variety of courses. Sometimes students have a hard time understanding De Morgan's Laws. In this activity, the students become sets and physically observe which of their classmates are in the complement of a set. They then compare different combinations of sets and discover the identities on their own.

Supplies Needed	Class Time Required	Group Size
1 set card per student*	15-20 minutes	Whole class

*Using cardstock, create *set cards*. A set card is a piece of paper with set names printed on one side. In large letters, print an "A" on one quarter of the cards, "AB" on another quarter, "B" on the third quarter, and leave the last quarter of the cards blank. Shuffle the cards together.

Running the Activity

Before beginning the activity, the instructor may want to discuss the set operations of union, intersection, and complement.

Begin the activity by giving each student a copy of the handout and one set card. Explain to the class that the whole class is the universal set for the discussion. Each student's set card indicates whether he or she belongs to subset A (A), to subset B (B), to both subsets A and B (AB), or to neither subset A nor subset B (blank). Each participant should stand up and take his or her handout, set card, and a pencil or pen with them.

Construct each of the sets in Table 5.3 in one part the room, and designate a different corner of the room as the set complement corner for those students not in the set being constructed. Figures 5.6a and 5.6b show students acting out their sets. After forming each set, use the table on the handout to record which set cards are in the set, and which are in the complement.

Table 5.3: Sets to Construct

$A \cap B$	$A \cup B$	$A^c \cap B$	$A^c \cup B$
$A \cap B^c$	$A \cup B^c$	$A^c \cap B^c$	$A^c \cup B^c$

Once all of these sets and their complements have been formed and the compositions recorded, have the students return to their seats and, in groups of 4-6, determine which of the sets are the same. Ask students to write their equalities on the board as shown in the top four lines of Figure 5.6c. Encourage the students to search for and identify patterns until they ultimately discover those shown in the the last three lines of Figure 5.6c. Explain that these patterns are known as De Morgan's laws, first proved by the British mathematician Augustus De Morgan (1806-1871). Then have students finish the rest of the handout.

Suggestions and Pitfalls

This activity works best when every student is participating in the action. However, if you have a very large class, it can be done by choosing 10-15 students to be the actors and having the rest of the class help the actors, observe the results, and fill in the handout.

This activity easily translates to propositional logic if the instructor changes the language from set union, intersection, and complements to AND, OR, and negation.

To decrease the time required for this activity, you could have the students construct only the four sets on the first line of Table 5.3.

If there is enough space in the room, the instructor may choose to use tape, yarn, feather boas or any available material to mark a life-sized Venn diagram on the floor of the classroom. Students could then fill in the Venn diagram according to which set card they were given by standing in the appropriately marked location. Then, instead of having students move to different parts of the room depending on whether they are in a given set or not, students who are in the sets under discussion could simply raise their cards above their heads and students not in the set could sit down. This has the benefit of showing where the elements of the sets in question lie within a Venn diagram.

(a) The set $A^c \cup B^c$.

(b) The set $(A^c \cup B^c)^c$.

(c) The set equalities and rules for applying De Morgan's Laws.

Figure 5.6: The class forming sets and discovering the rules.

Living DeMorgan's Laws – Class Handout

You are on the set construction crew today! After constructing each set, record who is in the set and who is in its complement in the following table. For example, if your instructor asks you to form the set $A \cap B$, those who have Card AB will be in the set, and all others will be in the complement; these results are recorded in the first two lines of the table.

Students in Each Set

Set	Card A	Card B	Card AB	Blank Card
$A \cap B$	no	no	yes	no
$(A \cap B)^c$	yes	yes	no	yes
$A \cup B$				
$(A \cup B)^c$				
$A^c \cap B$				
$(A^c \cap B)^c$				
$A^c \cup B$				
$(A^c \cup B)^c$				
$A \cap B^c$				
$(A \cap B^c)^c$				
$A \cup B^c$				
$(A \cup B^c)^c$				
$A^c \cap B^c$				
$(A^c \cap B^c)^c$				
$A^c \cup B^c$				
$(A^c \cup B^c)^c$				

Two sets are equal if they have exactly the same elements. In this case, two sets are equal if and only if they are composed of students with the same types of cards. Compare the sets in the chart above and find all pairs of equal sets.

Finish the sentence: The complement of the intersection of two sets is equal to

Finish the sentence: The complement of the union of two sets is equal to

5.10 Using Circuits to Teach Truth Tables

Jenna P. Carpenter, Campbell University

Concepts Taught: logical statements, AND and OR statements, truth tables, circuits

Activity Overview

This activity, which addresses basic logical statements, is appropriate to use in a wide variety of courses ranging from math for liberal arts to real analysis. Logic can be mistakingly perceived as a theoretical topic without much application to the "here and now". This one-day in-class lab using series and parallel circuits and a light bulb helps teach the concepts of AND and OR statements and illustrates an important application. The visual and hands-on nature of this activity provides an opportunity to accelerate student understanding of both circuits and logic statements; most students, even those with no background in circuits, are able to translate between these two representations with ease. In addition, the activity helps reinforce that the concepts in logic are relevant to real world applications, such as the circuits found in cell phones and computers.

Supplies Needed (per group)	Class Time Required	Group Size
1 C battery 8 alligator clip test leads* † 1 C battery case* † 1 small light bulb with wires* † 6 large, metal paper clips † 1 large clear, plastic, resealable bag †	1 class period	2-4 students

*These materials can be ordered from RadioShack or a similar store.
† Each group will need a circuit kit (Figure 5.8a) consisting of the leads, the battery case, the light bulb, and the paper clips all stored in the plastic bag. Each group will also need a battery, but this item should be stored separately, for safety purposes.

Running the Activity

This activity runs best if you have already covered AND and OR statements, as well as truth tables. Start by giving each group one circuit kit and one battery. Let the students know that one direct application of AND and OR statements is that of series and parallel circuits. The concepts are identical; we just relabel the items. When you flip on a switch (like a light switch), it closes or connects the circuit, turning on the light; when the switch is off, the circuit is open, so no current is flowing, and the light does not illuminate. Draw a simple circuit diagram on the board to show this (Figure 5.7). Have students assemble a simple circuit with their kits. Then have them demonstrate that when the switch is open the bulb does not light and when it is closed, the bulb lights.

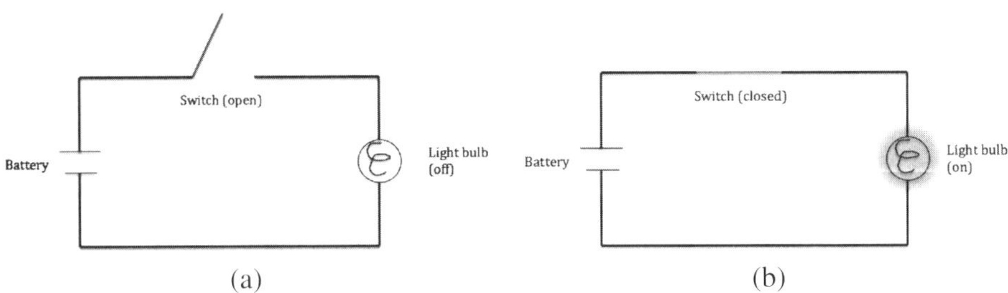

Figure 5.7: (a) A circuit that is off. (b) A circuit that is on.

Explain that a *series circuit* (Figure 5.8b) has two switches in series, one after the other. Both have to be closed in order for the circuit to be complete and the light to illuminate. Draw a series circuit. Have students assemble a series circuit with their kits and demonstrate when the bulb lights. A *parallel circuit* (Figure 5.8c) has two switches in parallel, so the current has two paths to follow. As long as one of the switches is closed, the circuit is completed, and the light illuminates. It also works if if both of the switches are closed, but not required. Draw a parallel circuit. Have the students assemble a parallel circuit with their kits and demonstrate when the bulb lights. Explain the analogy. Statements P and Q are switches. Statement P is true when switch P is closed. Statement P is false when switch P is open. A series circuit is the same as an AND statement. Both statements (switches) must be true (closed) for the entire statement (circuit) to be true (the light bulb lights). Similarly, a parallel circuit is the same as an OR statement. As long as at least one of statements (switches) is true (closed), the entire statement (circuit) is true (the light bulb lights).

Now write a more complicated compound AND/OR statement and have the students draw the analogous circuit, then assemble it with their kits. Complete the truth table for the statement and have students illustrate several of the rows of the truth table with their kits, noting that the overall statement being true corresponds to the bulb lighting. Then, draw a more complex series and parallel circuit on the board and have students write down the analogous AND/OR statement and work out the truth table for the statement.

Continue with increasingly complex examples. Once students are comfortable with these ideas, pass out the handout. Students can construct the analogous circuit for each problem on the handout and compare the truth table values to the instances when the light bulb lights.

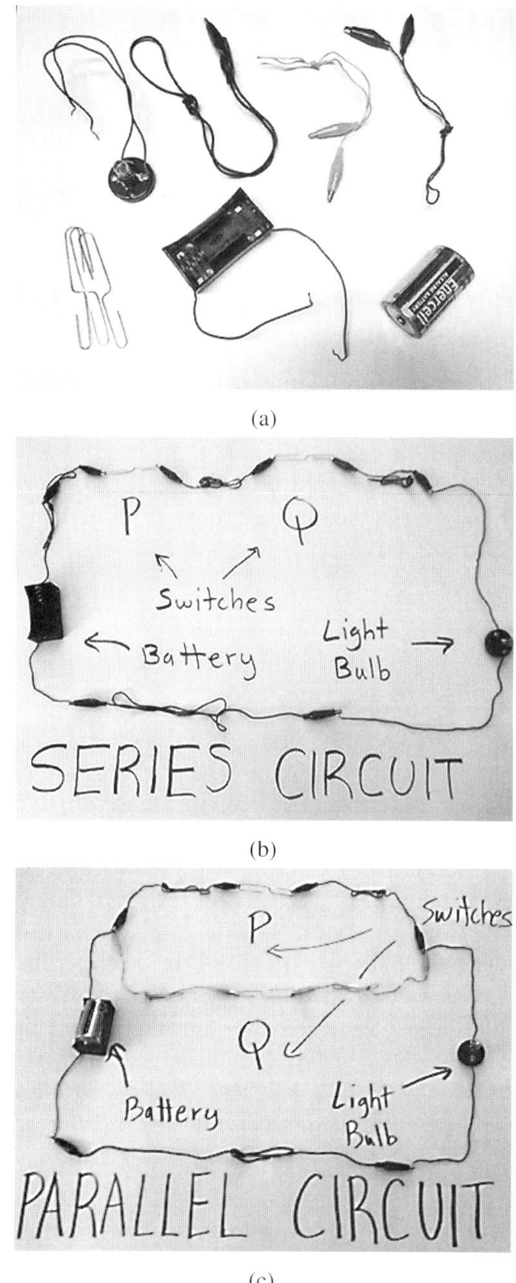

(a)

(b)

(c)

Figure 5.8: (a) A circuit kit. (b) A series circuit. (c) A parallel circuit.

Suggestions and Pitfalls

If you ask your students to construct more complicated circuits than those provided in the handout, be sure each kit is stocked with a sufficient number of additional paper clips and alligator clip test leads.

Have students rotate who is the lead assembler of the circuits so that every student has an opportunity to create circuits.

Make sure that batteries that you purchase are powerful enough to make the bulb light up brightly. The light can be hard to see in a bright room, particularly if the battery is not powerful enough. If the bulb doesn't light up brightly, have students cup their hand over it to see that it is lit.

Using Circuits to Teach Truth Tables – Class Handout

1. For each logic statement, draw the analogous circuit diagram and construct and complete the corresponding truth table. Construct each of these circuits and use them to check the values in your truth tables.

   (a) $(P \vee Q) \wedge (R \vee (S \wedge T))$

   (b) $((P \wedge Q) \vee (R \wedge S)) \vee (T \wedge U)$

   (c) $(P \wedge (Q \vee R)) \vee (S \wedge T)$

2. For each circuit, give the analogous logic statement and construct and complete the corresponding truth table. Construct each of these circuits and use them to check the values in your truth tables.

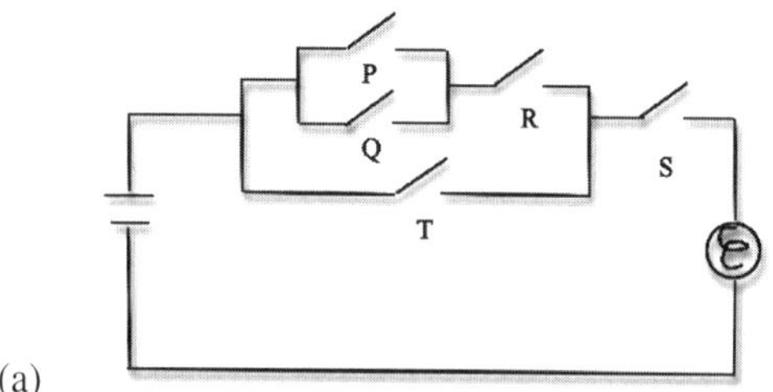

(a)

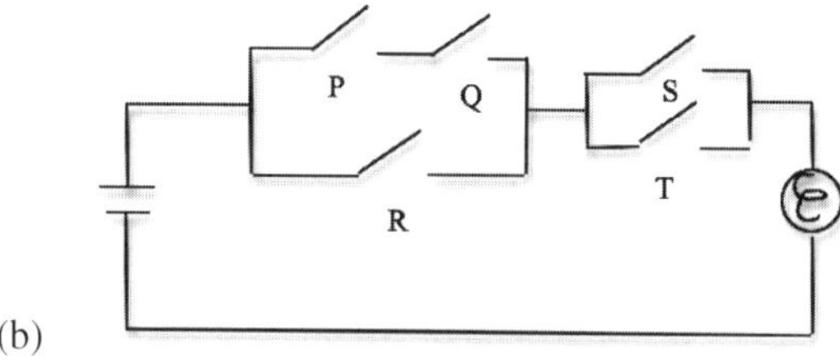

(b)

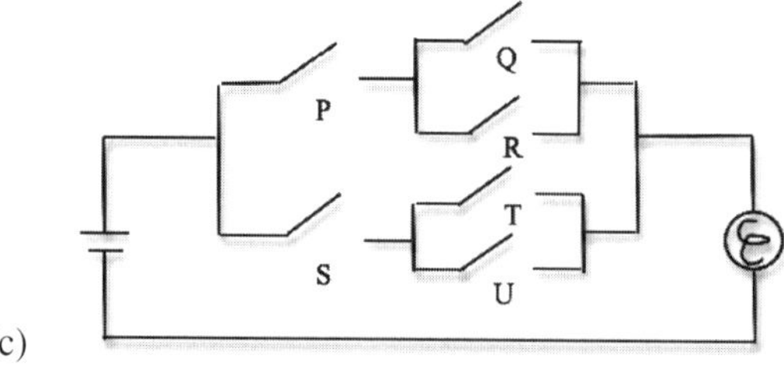

(c)

5.11 Determining the Validity of an Argument Using True/False Cards

Robert Franzosa, University of Maine

Concepts Taught: logical statements, truth tables, valid and invalid arguments

Activity Overview

In a logic or introduction to abstract mathematics course, students typically examine the validity of arguments by completing and analyzing truth tables. This activity provides an alternate approach where the class works collaboratively as a deductive logic computer using True/False cards. The class will examine deductive arguments and determine whether or not they are valid. The assumptions in an argument will be addressed successively, after which conclusions will be discussed.

Supplies Needed	Class Time Required	Group Size
A set of True/False cards*	50 minutes	2-3 students

*The set of True/False cards consists of 32 cards, each approximately $8\frac{1}{2}'' \times 11''$, and each marked with a unique distribution of True or False for five variables A, B, C, D, and E. See Figure 5.9a.

Running the Activity

To begin, provide each student with a copy of the handout and distribute all the cards evenly across the class. Depending on the size of your class, some students might receive multiple cards; in a large class, a group of students may need to share a card. Have students work in groups of 2-3 to discuss the cards that they have while the class works through the arguments.

As a class, the students will use the True/False cards and deductive reasoning to work through the problems in the handout. For example, for Problem 1a in the handout, have students consider the modus ponens argument which states that given the assumption *If A, then B* and the assumption A, we may infer B:

$$\begin{array}{l} \textit{If } A\textit{, then } B \\ \underline{A \qquad\qquad\qquad} \\ B \qquad\qquad\qquad . \end{array}$$

Have the class hold up the cards on which *If A, then B* is true. Then tell the class to continue holding up the cards on which A is also true, but put down the cards on which A is false. At this point, the cards that are being held up should be the cards on which both *If A, then B* and A are true. The resulting cards are shown in Figure 5.9b. (With more assumptions in an argument, continue the process, considering each assumption one by one.)

Next, collect the cards that remain after all assumptions are addressed and show them to the class. Then discuss the conclusion. In our example, B is true on all of the cards that remain and therefore you can state that B is a valid conclusion to draw from the given assumptions.

As a class, have students continue working through the handout. In Problems 1b and 1c, students encounter an invalid argument. In Problem 1c, they will also encounter negations. Explicitly, in Problem 1c, students consider the invalid denying-the-antecedent argument:

$$\begin{array}{l} \textit{If } A, \textit{ then } B \\ \underline{\sim A \qquad\qquad} \\ \sim B \qquad\quad . \end{array}$$

As in the previous example, begin by having the students hold up all the cards on which *If A, then B* is true. Next, you want the students to continue holding up the cards on which $\sim A$ is true, putting down the rest. Of course $\sim A$ is true if and only if A is false, so the cards that should remain raised are those on which *If A, then B* is true and A is false. The resulting cards are shown in Figure 5.9c.

Rather than collecting all of the cards that remain raised (16 of the 32 cards), have the students look around at the cards that are raised. Point out that $\sim B$ is true on some of them but not all of them. Therefore, $\sim B$ is not a valid conclusion to draw from the assumptions.

In a similar manner, work through the rest of the handout. The exercises in Problem 1 address many of the important logic structures needed for proofs, such as syllogism, cases, and contradiction. Problems 2 and 3 are couched in the form of a logic puzzle but can be solved using the same cards. Problem 2 presents a conclusion and asks students to discuss whether or not it is valid; this gives students the practice they need in walking through the logic for such a word problem. Alternatively, Problem 3 presents just the assumptions, and students use the cards to discuss what conclusion can be drawn; specifically, they need to recover who is going to the beach (Carl and Dan).

Suggestions and Pitfalls

It is necessary that the students understand the conditions under which an *If-then* statement is true or false. You may want to do a quick review of this information before starting the activity. Also, mistakes are likely to occur when students evaluate whether or not an assumption is true for a card that they have. Having the students work in pairs can help to reduce such mistakes. Also, by numbering the cards and knowing which numbered cards should remain after addressing the assumptions in an argument, you will have a means for a quick check to make sure that the correct cards resulted.

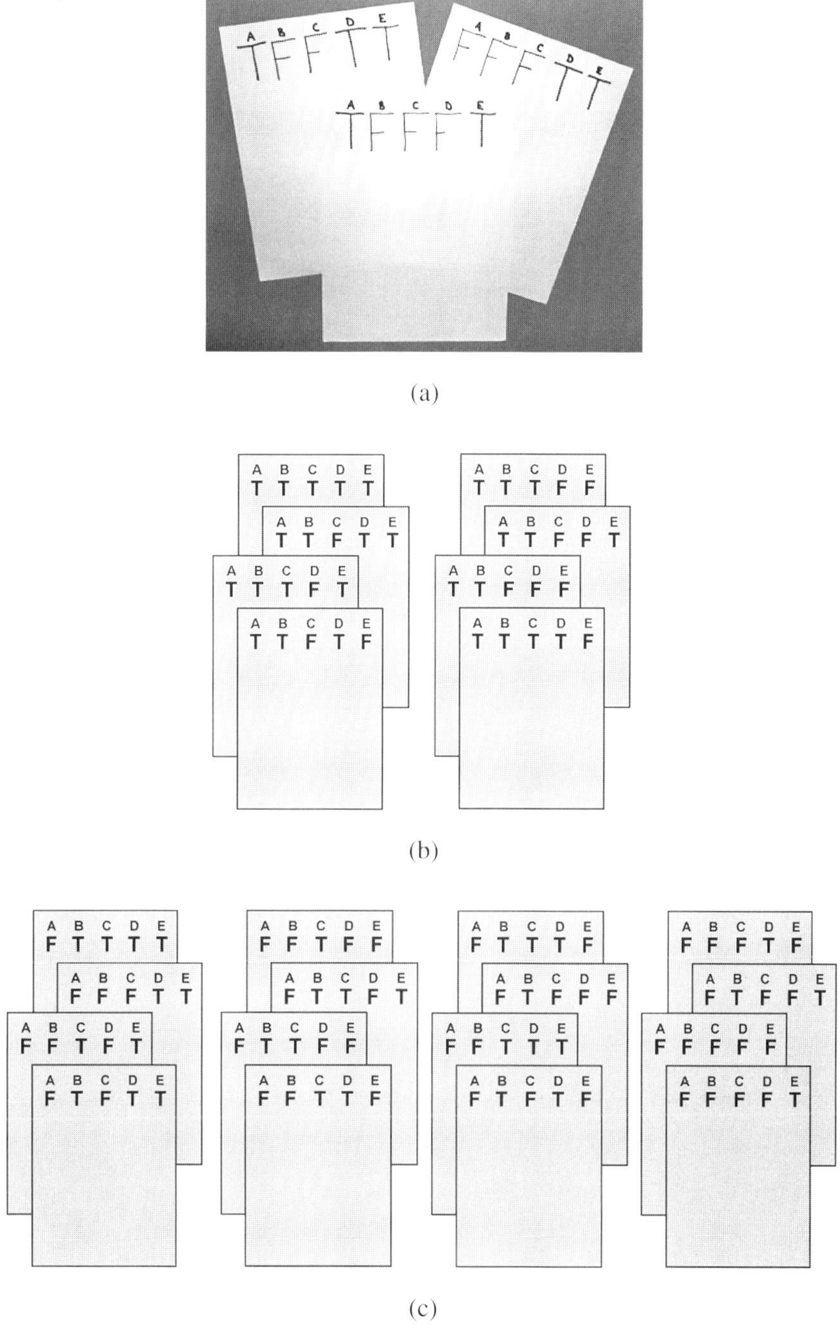

(a) (b) (c)

Figure 5.9: (a) Example of True/False cards. (b) True/False cards on which both the statement *If* A, *then* B and the statement A are true. (c) True/False cards on which both the statement *If* A, *then* B and the statement $\sim A$ are true.

Determining the Validity of an Argument – Class Handout

Address the following arguments and logic puzzles using the True/False cards:

1. Some basic arguments. In each case we will determine whether or not the argument is valid:

a. *Modus Ponens* *If A, then B* *A* *B*	b. *Assuming the Consequent* *If A, then B* *B* *A*	c. *Denying the Antecedent* *If A, then B* $\sim A$ $\sim B$	d. *Hypothetical Syllogism* *If A, then B* *If B, then C* *If A, then C*
e. *Affirming a Disjunct* *A or B* *A* $\sim B$	f. *Proof by Cases* *A or B or C* *If A, then D* *If B, then D* *If C, then D* *D*	g. *Contradiction Example* *If* $\sim A$*, then B* *If* $\sim A$*, then* $\sim B$ *A*	h. *Disjunctive Syllogism* *A or B* $\sim A$ *B*

2. Show that we can conclude that Eddie and Carrie passed their logic course given these statements.

   If Carrie failed, then Annie passed.
   It is not the case that Carrie passed and Eddie failed.
   If Annie passed, then Carrie and Eddie also passed.
   (Let A represent "Annie passed", C represent "Carrie passed", E represent "Eddie passed".)

3. Who is going to the beach?

   If Alyssa is going to the beach, then so is Bonnie.
   Alyssa and Carl are going to the beach if Erin is.
   Either Bonnie or Dan is going to the beach, but not both.
   Carl is going to the beach if Dan is, otherwise he isn't.
   If Carl isn't going to the beach, then Erin is.

5.12 Proof Rearrangements

Julie Barnes, Western Carolina University

Concepts Taught: proof organization

Activity Overview

When students are first introduced to the idea of writing a proof on their own, they often have difficulty determining the order in which it must be written. This is particularly true for proofs involving quantifiers, but can be an issue with any logical argument. This activity provides students with an opportunity to practice writing proofs in logical order by arranging puzzle pieces comprised of parts of a proof. It can be used in an introduction to proof class, where students arrange new puzzles as different techniques are introduced. Alternatively, this activity can also be used as a quick review of proving techniques on the first day of an upper level class like real analysis or abstract algebra.

Supplies Needed (per group)	Class Time Required	Group Size
Prepared puzzle*	10-15 minutes	2-3 students

*You will need to make the puzzles before class. Choose one of the puzzles provided on pages 213 and 214. Photocopy the page (preferably on card stock), and cut along the lines with scissors to prepare one complete puzzle. You will need one complete puzzle for each group. When you create the puzzle pieces, since each group will be working on the same proof, it is helpful to make each complete set of puzzle pieces a different color. That is, one group will have an entire set of blue pieces, another will have an entire set of red, etc. This helps keep the puzzles organized.

Running the Activity

In class, provide each group of students a complete puzzle and have them work together to arrange the proof in a correct order as seen in Figure 5.10. Sometimes more than one proof order is logically correct. As students work, walk around the room answering questions, and let students know when they have completed the activity correctly. You can give hints by asking them how the proof should start or by telling them how many cards are out of place. Once everyone has completed their puzzles, bring everyone together for a class discussion. Have students identify any parts that they found to be difficult and have them share suggestions on how to determine the correct order of the proof.

Figure 5.10: Students working on rearranging a proof.

Suggestions and Pitfalls

If you intend to use the pieces again in the future, ask the students to shuffle them before handing them back to you so the puzzles are ready for the next use.

The idea behind this activity could be modified for any upper level class. Simply take a short proof that students find difficult and turn it into a puzzle. This idea also works well for helping students distinguish between two similar but different definitions. For example, to help students appreciate the proper use of quantifiers in the definitions of continuity and uniform continuity, you can create a single puzzle with pieces of both definitions; students then have to sift through the pieces to obtain both correct definitions.

If your students need more of a challenge, you can add extraneous cards, and students will have to determine which cards are not necessary. You can also include blank cards for them to add a missing component of the proof. In addition, you may want to have extra, more difficult puzzles available for any groups that finish more quickly than expected.

If you have space in your classroom, you could create large puzzles by rewriting the information from each box on a separate piece of card stock with a marker. For large pieces, students need to spread out on the floor or use large tables; this forces students to work as a team. See Figure 5.11 for photographs of students arranging a large proof puzzle.

Figure 5.11: Students working on rearranging a proof that is written on larger cards.

Sample Proof Puzzle Pieces – Proof 1

Below are puzzle pieces for a proof that the sum of two even numbers is even.

$2n + 2m$	Since $n + m$ is an integer,	$=$
$p + q$ is even	such that $p = 2n$ and $q = 2m$.	$p + q$
$2(n + m)$.	Let p and q be even numbers.	$=$
n and m	Then, there exist integers	Then,

Solution: Let p and q be even numbers. Then, there exist integers n and m such that $p = 2n$ and $q = 2m$. Then, $p + q = 2n + 2m = 2(n + m)$. Since $n + m$ is an integer, $p + q$ is even.

Sample Proof Puzzle Pieces – Proof 2

This page provides puzzle pieces for a proof that $\lim_{x\to 2} 3x + 1 = 7$.

$\lvert 3x - 6\rvert$	ϵ	$<$
if $0 < \lvert x - 2\rvert < \delta$,	$\frac{3\epsilon}{3}$	$=$
$3\lvert x - 2\rvert$	Let $\epsilon > 0$.	$=$
For all x,	Let $\delta = \frac{\epsilon}{3}$.	$=$
then $\lvert(3x + 1) - 7\rvert$	3δ	$=$

Solution: Let $\epsilon > 0$. Let $\delta = \frac{\epsilon}{3}$. For all x, if $0 < |x-2| < \delta$, then $|(3x+1) - 7| = |3x - 6|$ $= 3|x - 2| < 3\delta = \frac{3\epsilon}{3} = \epsilon$.

5.13 Properties of Functions on Finite Sets Using Candy

Erin Bancroft, Grove City College

Concepts Taught: one-to-one, onto, finite sets, order of a set

Activity Overview

When students transition from the calculus sequence to upper level math courses, they often take an introduction to proof course in which they learn about the building blocks of mathematics: sets and functions. In the calculus sequence, students deal with functions on infinite sets such as the real numbers, but they typically don't have much context for working with finite sets and functions between finite sets. This activity gives students a hands-on experience using candy to illustrate the definitions and properties of functions between finite sets. It also introduces and solidifies the formal definitions of one-to-one, onto, and bijective. The tactile experience of physically touching and connecting input and output with candy and toothpicks seems to provide a stronger memory foundation for the definitions than just drawing the diagrams. Often students recreate the specific examples done in the activity and use them to try to re-formulate the definitions when they get stuck.

Supplies Needed (per group)	Class Time Required	Group Size
1 snack size bag of Skittles®* 1 snack size bag of regular M&M's®* 6 toothpicks	35-40 minutes	1-4 students

*Non-edible objects or alternative candies can be used in place of the standard 5-colored Skittles and the standard 6-colored M&M's, but it is helpful if the objects come in a variety of colors (at least four).

Running the Activity

To start this activity, divide students into groups and give each group a packet of Skittles, a packet of M&M's, and six toothpicks. Begin by reviewing or introducing the definitions of function, domain, codomain, and range. Have students open the candy and take out one of each color from each bag. If a student is missing a color of Skittles or M&M's, have them get it from another student. Tell students that they will be creating functions between two finite sets where the set S is the domain, and the set M is the codomain. The set S will consist of the five different colored Skittles, and the set M will consist of the six different colored M&M's. Have students line up the elements in the domain and codomain vertically

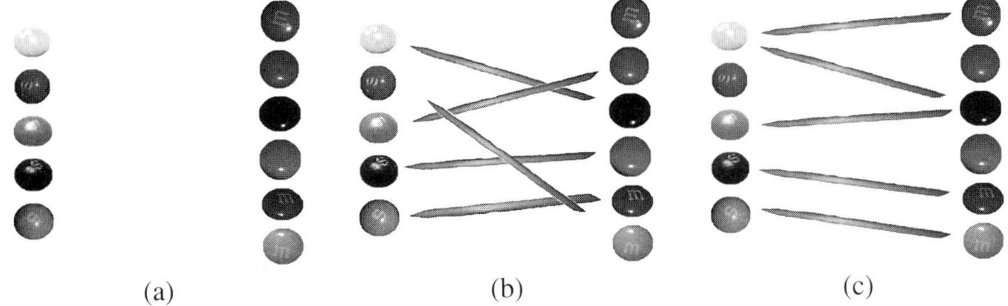

Figure 5.12: (a) Setting up candy. (b) An example of a function. (c) An example of something that is not a function.

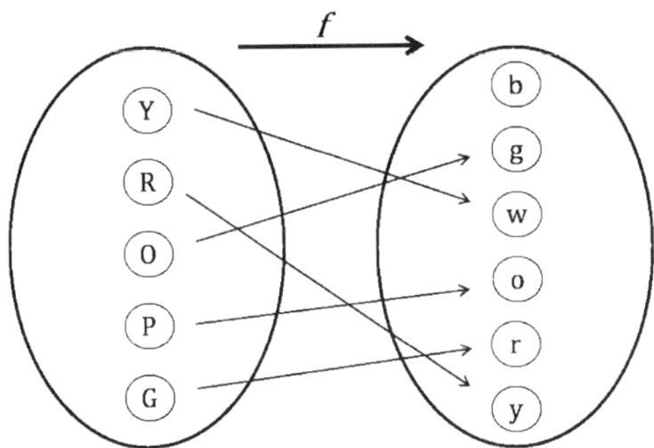

Figure 5.13: Diagramming the function from S to M represented in Figure 5.12b.

in front of them with the codomain a toothpick's length to the right of the domain. See Figure 5.12 as an example.

Ask groups to create a function. For each element in the set S, students should place a toothpick such that it points from one of the Skittles to one of the M&M's. Have them look at the relations created by other groups around them and verify that they are functions. While students do this, pass out the handout, and help to correct any student examples that are not functions. Pick the function of one of the students and demonstrate to the entire class how to diagram that function; a sample diagram is shown in Figure 5.13. Also demonstrate diagramming an example that is not a function, possibly one chosen from the class; have a volunteer describe the reasons why this second example is not a function.

Students should then work through the handout in their groups. In the handout, students are presented with the definitions of surjection, injection, and bijection and are asked to create and diagram functions from Skittles to M&M's that satisfy the definitions. They will also investigate the relationship between the orders of the domain and codomain in functions that are surjective, injective, and bijective. While students are working, walk around the room to correct errors, answer questions, and assist students in diagramming their functions correctly. After students have completed the handout, discuss the solutions briefly as a class to be sure that each group has a set of correct diagrams and conclusions.

Suggestions and Pitfalls

Students will often try to use all of one color from S or M for the domain or the codomain, so be sure to emphasize that the domain and codomain set should consist of one candy of each color. When choosing a student's example of a function to use as an example for the class, try to find a function that is not one-to-one as this helps to contrast multiple elements in the domain going to the same element in the codomain with an element in the domain going to multiple elements in the codomain. Students may also have difficulty realizing that not mapping an element of the domain to an element of the codomain constitutes a relation that is not a function as they are used to the domain being defined as only the values that "work" in a function.

For further study, a discussion of how to prove the results of Problem 7 on the handout leads nicely into a lesson on the pigeonhole principle. Alternatively, this activity could also precede a lesson on formally proving that a function is surjective, injective, or bijective.

Properties of Functions on Finite Sets Using Candy – Class Handout

S = the set of different colored Skittles® (one of each of the five colors)

M= the set of different colored M&M's® (one of each of the six colors)

1. Draw a diagram of the function $f : S \to M$ that you created. What are the *orders* (number of distinct elements) of the domain and the codomain? Use $|S|$ to denote the order of the set S and $|M|$ for the order of M. Put a box around the range of the function.

2. Draw a diagram of a relation from S to M that is not a function and write down how it violates the definition of a function. Draw a second diagram of a relation from S to M that is not a function for a different reason.

3. A function f is said to be *onto* or *surjective* if for each element y in the codomain, there exists an element x in the domain such that $f(x) = y$. In other words, the range is equal to the codomain. Let Y be the set consisting of the red, blue, green and yellow M&M's. Create an onto function from S to Y. Create a second function that is not onto from S to Y. Draw the diagrams of both.

4. A function f is said to be *one-to-one* or *injective* if for all elements x_1 and x_2 in the domain, if $f(x_1) = f(x_2)$ then $x_1 = x_2$. In other words, every element in the domain maps to a unique element in the codomain. Create a function from S to M that is one-to-one. Create a second function from S to M that is not one-to-one. Draw the diagrams of both.

5. A function is said to be a *bijection* or *bijective* if it is both one-to-one and onto. Define a domain A and a codomain B consisting of Skittles and M&M's respectively, and then create a function that is a bijection from A to B. Draw the diagram.

6. Is it possible to define a function from M to S that is a bijection? Why or why not?

7. What can you say about the orders of the domain and codomain when you know that a function is onto, one-to-one, or a bijection? Does one have to be larger than the other? Can they be equal? Or is there not enough information to decide? Draw some examples below, and when you have a conjecture, think about how you might prove it is true.

   If a function $f : A \to B$ is onto, then $|A|$? $|B|$. That is, how is the size of A related to B?
   If a function $f : A \to B$ is one-to-one, then $|A|$? $|B|$.
   If a function $f : A \to B$ is bijective, then $|A|$? $|B|$.

5.14 SET® in Combinatorics/Discrete Math

Elizabeth McMahon, Lafayette College

Concepts Taught: combinations, multiplication principle, combinatorics

Activity Overview

This activity, which introduces students to the game of SET and counting techniques, can be used in a combinatorics or a discrete mathematics course. After students develop a sense for what groups of three cards meet the criteria for a *set*, they can use combinatorial techniques to count the number of sets satisfying various additional criteria. The solutions to the questions addressed in this activity will use the multiplication principle (if you have a pair of independent choices to make, and there are m choices for the first and n choices for the second, then there are mn choices in total), combinations (counting the number of sets based on the selection of a subset of attributes), and some interesting facts about the game of SET. Most students really enjoy the game, giving them motivation to apply techniques from class to answer natural questions that arise about the deck.

Supplies Needed	Class Time Required	Group Size
Several decks of SET cards (it helps for each group to have a third of a deck) Document camera (optional)	˜30 minutes, depending on how much time is spent learning the game	4-8 students*

*If some students know the game well, consider putting them together. Grouping students who already know the game allows them to complete the activity without intimidating those who are learning.

Running the Activity

If the students have not had any previous exposure to the game of SET, begin by describing the deck, the cards, and what constitutes a *SET*. The SET cards have symbols characterized by four attributes: number, color, shading, and shape (writing those on the board can be helpful). A *SET* is three cards where, for each attribute, each incidence of that attribute is all the same or all different. The SET website [1] has tutorials on how to play.

If you have access to a document camera projection system, you can put cards on that to illustrate *SETs* and non-*SETs*. Show an example where three attributes are the same and one is different, one where two are the same and two are different, one where one is the same and three are different, and one where all are different. Such examples are pictured in Figure 5.14a. Then show some examples of non-*SETs* and have students explain why they are not *SETs*. Several examples are shown in Figure 5.14b. It can be useful to tell your students that if you have three cards where you can say, "Two cards are X but the other

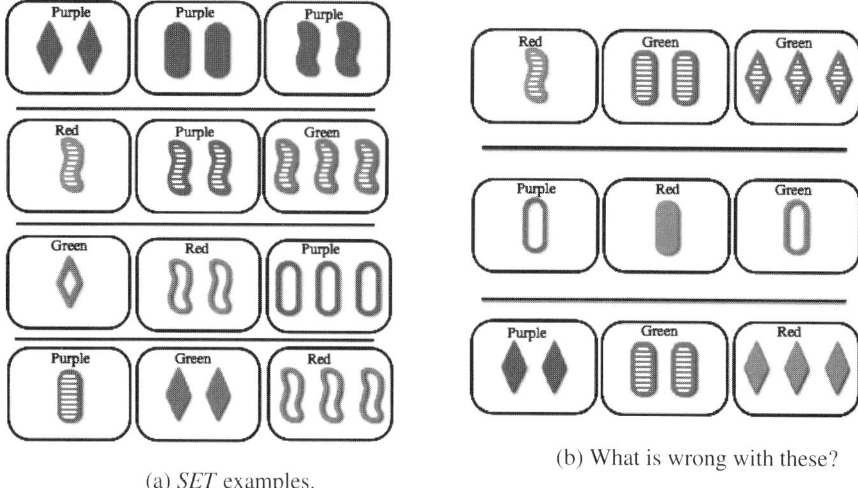

(a) *SET* examples.

(b) What is wrong with these?

Figure 5.14: Examples of (a) *SETs* and (b) non-*SETs*.

is not," then it is not a *SET*. Also point out that given any two cards, the third card that completes the *SET* is uniquely determined.

Once most students are able to identify *SETs*, put them into groups, and pass out the handout. Let the groups start by playing one round of the game, even if it is not with a complete deck. Have students begin working on the handout while experimenting with the deck. As you move through the room, make sure that students understand the rules for how to construct *SETs*, and provide assistance as students work on the handout. If students do not finish the handout during class, they can complete it as an out of class assignment.

Suggestions and Pitfalls

If students have not played SET before, you can introduce the game prior to the day of the activity and assign students the task of going to the SET website [1] and playing the Puzzle of the Day. This is the kind of assignment students typically enjoy. The website reports how long it took to complete the puzzle; students could write down their time for you to collect. If there are two days before class, have them do it twice and compare their times for the two puzzles. This often leads to an interesting observation: some puzzles seem to be harder than others. What characterizes the easier ones? It might have to do with how many attributes the *SETs* share, which is the subject of the handout.

One possible pitfall with this activity is that students can get distracted by other aspects of the game; of course, that is not really a pitfall. If they do, you could organize a tournament, or get your Math Club involved. SET is a very mathematical game, and there are many more questions you can explore with it. The game can also be used in geometry, linear algebra, and liberal arts math classes.

For a stronger class or a class where you have more time, you may want to give this additional guidance for Problem 5 on the handout:

> In n-attribute SET, the number of *SETs* where all but one attributes are the same ought to be the least common. Prove that it is always least common, independent of the value of n. Prove that the number of *SETs* with k attributes the same is *unimodal*: it increases as k increases to some value, then decreases after that point. Then, find the maximum as a function of n.

References

1. SET Enterprises, www.setgame.com.

2. SET © 1988, 1991 Cannei, LLC. All rights reserved. SET® and all associated logos and tag lines are registered trademarks of Cannei, LLC. Used with permission from Set Enterprises, Inc.

3. E. McMahon, G. Gordon, R. Gordon, and H. Gordon, The Joy of SET: The Many Mathematical Dimensions of a Seemingly Simple Game, *Princeton University Press,* November 2016.

SET® in Combinatorics/Discrete Math – Class Handout

- SET is played with a special deck of cards, each of which has symbols characterized by four attributes:

  Number: 1, 2, or 3 symbols. **Color**: Red, purple, or green.
  Shading: Empty, striped, or solid. **Shape**: Ovals, diamonds, or squiggles.

  A *SET* is three cards where, for each attribute, each incidence of that attribute is all the same or all different.

- Two important notes: (1) The number of attributes that are the same can vary. (2) If you have three cards where there is any aspect where you can say "Two are X but one is not," then the three cards are not a it SET. For example, if two are ovals but one is not, it is not a *SET*. If exactly two are striped, it is not a *SET*.

- To play: 12 cards are laid on the table. The first person to spot a *SET* yells "*SET!*", and takes the cards that make a *SET*, and those cards are replaced. If at some point, everyone agrees that there aren't any *SETs* on the table, three more cards are added, but if a *SET* gets taken, don't replace the cards unless there's no *SET* again. When there are no more cards in the deck and no *SETs* on the table, the winner is the one with the most *SETs*.

1. Quick questions:

   (a) How many cards are there in the whole deck?

   (b) How many *SETs* are there in the whole deck?

   (c) Consider the card with two red solid squiggles. How many *SETs* contain that card? Is the answer unique to that card, or does each card belong to that many *SETs*? Justify your answer.

2. When you lay out the first 12 cards, how many *SETs* do you expect to see on average? *Hint: What is the number of ways of choosing three cards? The probability that three cards form a SET is the fraction (number of SETs)/(number of ways of choosing three cards). Now what do you do?*

3. The questions below ask for the number of *SETs* with various properties. Since you know how many *SETs* are in the deck, also figure out what percentage (or fraction if you prefer) of *SETs* has those properties.
   (a) What is the number of *SETs* with all attributes different?
   (b) What is the number of *SETs* with one attribute the same and three different?
   (c) What is the number of *SETs* with two attributes the same and two different?
   (d) What is the number of *SETs* with three attribute the same and one different?

4. A standard SET deck has cards with four different attributes, but you can make a version with more or less attributes. For example, you can have SET with only three attributes by taking all the solid cards, or you can have SET with five attributes by taking three decks and decorating the backgrounds of two of the decks.
   (a) For 3-attribute SET, what is the number of *SETs* with all attributes different? With two attributes different and one the same? With one attribute different and two the same? What fraction or percentage of the *SETs* has those properties?
   (b) Answer the same questions for 5-attribute SET. (Pay attention: there are more possibilities.)

5. Summarize your results. How might you generalize these results?

5.15 Human-Powered Computing

Karen Bliss, Virginia Military Institute
Jessica Libertini, Virginia Military Institute

Concepts Taught: scripting, programming, algorithmic thinking

Activity Overview

While many students like to view themselves as tech savvy, the technologies to which they are most accustomed are designed with friendly interfaces that can be understood by non-sophisticated users. Many students lack experience in developing their own code or script, and are not competent in troubleshooting errors in their own code. This activity can be used to help students understand reasons their code might fail to run, why it might produce unexpected results, and how to interpret and fix error messages.

In an introductory numerical analysis or scientific computing course, common student coding difficulties include a failure to define and/or assign a value to a variable before it is used, a weak understanding of for loops and while loops, and confusion about language or software-specific notation. This specific activity and accompanying code target the first issue; however, it could be modified to address any of these concerns, and others as well.

In an effort to help students become self-sufficient in identifying problems with their code, students are each given a single line of a problematic code. Then, one at a time, they must execute their line of code at the chalkboard based only on the information that is already available from the previous students identifying bugs as they go. After manually executing some pre-bugged code, students try to run it on a computer and explore the meaning of the resulting errors.

Supplies Needed	Class Time Required	Group Size
Printed and cut code* Access to software or hardware (optional)**	5-10 minutes per code 15 minutes for handout	Class demonstration

*Samples of pre-bugged code are provided in Figure 5.15. Print these, including the line numbers, and cut them into strips such that each student can be given one numbered line of code.

**To complete the activity, students will need to enter the code into their computers so they can become familiar with the meaning of common error messages; this can be done outside of class.

Running the Activity

Give each line of code to a different student, and then have these students come to the board one at a time, in the order indicated by the line number for the code. Ask each student to

Code A: Defining Variables	Code B: Defining Variables in Loops
A Line 01 `x=3`	B Line 01 `n=3`
A Line 02 `y=4`	B Line 02 `for i=1:n`
A Line 03 `A=x+y`	B Line 03 `    b=i^2`
A Line 04 `c=a+x`	B Line 04 `    a=a+b`
A Line 05 `d=A+B`	B Line 05 `    i=i+1`
A Line 06 `B=-4`	B Line 06 `end`
A Line 07 `F=A+B+c+d`	B Line 07 `a`

Figure 5.15: Samples of code, each containing bugs.

execute the given command on his/her strip of paper, writing and/or erasing results on the board accordingly. For example, the student for Line 01 of Code A in Figure 5.15 would write `x=3`. If a student doesn't know how to execute his/her command, perhaps due to a bug in the code, engage the class in a discussion about why the student can or cannot execute the command as given. For example, in Line 05 of Code A, the variable `B` has not yet been defined. If the code contains a loop, as in Code B, make sure the students inside the loop understand that they will need to repeat their steps until the loop is complete.

In Code A, the main focus is on recognizing the importance of the order of defining and calling variables, as well as case sensitivity. Code B contains some common mistakes including reinitializing variables inside the loop, failing to define a variable, and manually iterating. With both codes, the *correct* version depends on what you want the code to do. As the class uncovers the bugs, lead the class in a discussion about the behavior of the code based on different debugged versions. For Code B, this discussion is particularly rich, as you can consider when one might want to define the variable `a`.

After going over the coding errors at the board, give your students the handout that contains the entire code from the activity. Either in class or outside of class, have them enter this code into their computers and attempt to execute it. The handout asks the students questions about the debugging process by identifying exactly what error messages they receive and decoding their meaning. The handout also includes questions to prompt students to think about some best practices in coding.

Suggestions and Pitfalls

It is best to run this activity after students have tried (and often failed) to write their own code. This activity gives them insights into the error messages they have been receiving and is designed to help them become more self-sufficient at debugging. Be sure you do not hand out the entire code before running the line-by-line activity, as students may be tempted to look ahead and use information out of sequence.

Some syntax errors, such as the case sensitivity in Line 04 of Code A, may not be caught during the at-the-board portion of the activity but will be addressed in the second part of the activity when the students enter the code you provide into their computers and attempt to execute it.

You are encouraged to make your own code designed to address concerns specific to your students, to your course, and to the language or software you are using, or to include elements from previous student submissions. For example, other common issues can include using spaces in variable names, trying to use reserved strings (such as e) as variable names, deciding what values need to be stored and how to store them, and preallocation of space for large matrices. As you build, cut, paste, and shape the code that you will use, it is advisable to target a small set of concerns. If you are tempted to address more issues than time permits during class, you may consider the alternative of having an out-of-class assignment in which students work through additional code by hand, showing each step while identifying and debugging any errors they uncover.

Using this human-powered computing approach is also valuable to explore the differences between for loops and while loops for code efficiency. Note that a thorough exploration of more advanced topics may require significantly more class time.

Human-Powered Computing – Class Handout

Code A: Defining Variables		Code B: Defining Variables in Loops	
Line 01	`x=3`	Line 01	`n=3`
Line 02	`y=4`	Line 02	`for i=1:n`
Line 03	`A=x+y`	Line 03	`    b=i^2`
Line 04	`c=a+x`	Line 04	`    a=a+b`
Line 05	`d=A+B`	Line 05	`    i=i+1`
Line 06	`B=-4`	Line 06	`end`
Line 07	`F=A+B+c+d`	Line 07	`a`

After attempting to execute the sample codes above on your computer, complete the following problems. Do not attempt to fix any errors until asked.

1. Are variable names dependent on capitalization?

2. What happens if your code calls a variable that has not yet been defined?

3. Does code run properly if you define a variable and then never use it? Is it a good idea? Explain.

4. In Code A, what is the first error message you receive? Write it down verbatim.

5. What do you think that first error message you received means? Fix it. Briefly explain how you modified the code to fix this issue.

6. Continue to debug Code A. What error messages do you get, and what do you do to fix them?

7. When you try to run Code B, you probably notice that you do not have a value for `a` in line 4. Suppose that we meant for `a` to start with a value of 2. You could address this by adding a new line before Line 02 that initializes `a` to 2, or by adding a new line before Line 04 initializing `a` to 2. Briefly explain the differences in these solutions.

8. Which of the following are acceptable variable names? If you find any that are unacceptable, explain why.
   `F, f, f1, f2*, f 4, function1`

9. Will you get an error message when running your code if you have two different variables, one named `function1` and another named `Function1` in the same code? Do you think this is a good idea? Explain your response.

Chapter 6

Junior / Senior Courses

6.1 Using Candy to Represent Equivalence Relations

Jenna P. Carpenter, Campbell University

Concepts Taught: equivalence relations

Activity Overview

Many textbooks present equivalence relations from an abstract viewpoint, even though equivalence relations are actually used in everyday life. This activity begins at the outset of the discussion on equivalence relations and provides students with a tactile and visual way to experience the concepts covered. Students use a bag of assorted pieces of candy to test, verify, and even discover concepts and theorems about equivalence relations.

Supplies Needed (per group)	Class Time Required	Group Size
Prepared bag of 10-12 hard candies *	15-20 minutes	1-4 students

*For each group, assemble an assortment of individually wrapped hard candies and place them in a resealable sandwich bag. It is best if you buy some candies that are all alike, such as a bag of peppermints and a bag of butterscotch, and some that contain different colors, such as a bag of Jolly Ranchers®.

Running the Activity

Pass out the bags of candy and the handouts prior to introducing the concept of equivalence relations. Let the students know that they will be using the candy to illustrate the concepts in this section.

Start by defining equivalence relations, and then have the students take their candy and make piles of candy so that each pile is alike in some way. Tell students that each piece of candy must be placed into some pile and that no piece of candy may belong to more than one pile. After they are finished, have each group explain their reasoning. Verbalize the criterion for two pieces of candy being equivalent, according to the way they grouped their candy. For example, if the criterion is that the candy had to be the same type, all peppermints would be grouped together, all butterscotch grouped together, and all Jolly Ranchers grouped together (Figure 6.1a). Another group may sort by color putting all of the red candy together, such as peppermints and red Jolly Ranchers (Figure 6.1b). If you have groups that sort the candy in different ways, emphasize that we can look at the same

(a) Candy grouped by type.

(b) Candy grouped by color.

Figure 6.1: Candy as it is rearranged.

objects (pieces of candy) and group them in different ways, depending on which attributes we want to emphasize (color, type of candy, etc.).

Have students use their candy to identify other terms as they are covered in class. For example, when introducing the term *equivalence class*, students should recognize that each pile of candy is an equivalence class. Similarly, when introducing the term *partition*, students should realize that the set of all of the piles is a partition of their bag of candy.

Have students use their candy to illustrate the properties of an equivalence relation:

> *reflexive*: every piece of candy is in the same pile as itself;
>
> *symmetric*: if candy A is in the same pile as candy B, then candy B is in the same pile as candy A;
>
> *transitive*: if candy A is in the same pile as candy B and candy B is in the same pile as candy C, then candy A is in the same pile as candy C.

To complete the activity, students will need to regroup their candy a different way at least two more times. Each time, have the students verbalize the criterion they used to group their candy.

By the time students see the theorem stating that candy can be partitioned into piles using any criterion and it will always be an equivalence relation, they will probably have already discovered this. That makes this usually tricky theorem seem intuitive. Also, while talking about the fact that the set of all equivalence relations on a set (such as the pieces of candy they have) and the set of partitions of the set of candy into piles are, in fact, equivalent, students will have discovered that, too. Again, it makes an abstract theorem clear and natural.

Suggestions and Pitfalls

As we move through the material, my students use the candy every time to illustrate each new theorem. You can have students complete each problem as you discuss the associated concepts, or wait and have students complete it at the end of the section.

If assigned as homework, students will need to assemble their own bag of candy or similar items to use in completing the problems.

Since it usually takes more than one day of class to discuss the material on equivalence relations, I find that this activity works best if you collect the bags of candy at the end of each class and bring them back the next class day. If you allow the students to keep the candy, it will not necessarily all show up each day.

When you have finished the topic, you may elect to let the students keep their candy, provided your institution has no policy against that. It is not a good idea to use chocolate because it may melt or candy that contains nuts given that some students may be severely allergic to nuts.

Using Candy to Represent Equivalence Relations – Class Handout

Use your bag of candy to complete the following activities. For each activity, there are one or more related problems, to help you connect the activity to the concepts about equivalence relations.

Part 1:
Organize your pieces of candy into piles so that the pieces of candy in each pile are alike in some way. Each piece of candy must be placed into some pile. No piece of candy can be in more than one pile.

1. Describe the criterion that you used to organize the candy.

2. Do the individual piles of candy satisfy the definition of an equivalence class? Explain why or why not.

3. Looking at all of the piles of candy collectively, do they satisfy the definition of an equivalence relation on your bag of candy? Explain why or why not.

4. Illustrate the properties of an equivalence relation (reflexive, symmetric, and transitive) using your grouping of the candy. Write out your responses.

5. Looking at all of the piles of candy collectively, do they satisfy the definition of a partition of your bag of candy? Explain why or why not.

Part 2:
Organize your candy into piles using a different set of criteria. As before, each piece of candy must be placed into some pile. No piece of candy can be in more than one pile.

6-8. Complete Problems 2-4 for this new arrangement.

Part 3:
Organize your candy again, using a set of criteria that is different from what you used the first two times. Again, each piece of candy must be placed into some group. No piece of candy can be in more than one pile.

9-11. Complete Problems 2-4 for this arrangement of your candy.

12. Reflecting on what you have learned from this exercise, explain the relationship between partitions and equivalence relations of a set.

6.2 Finding the GCD: Euclidean Disc Toss

Michael D. Smith, Lycoming College

Concepts Taught: Euclid's algorithm for finding the greatest common divisor

Activity Overview

Euclid's algorithm, typically encountered in abstract algebra and number theory, is a procedure used to compute the $\gcd(a, b)$, the greatest common divisor of two numbers a and b. This algorithm can be thought of as a "while" loop, where you perform the same sequence of steps (finding a quotient and a remainder) to pairs of successively smaller numbers. To underscore the recursive nature of Euclid's algorithm, we model it with a disc toss during which the students follow the same sequence of steps at each iteration of the algorithm.

Supplies Needed (per team of 3)	Class Time Required	Group Size
3 flying discs, such as Frisbees® 10 station markers* 60 small mailing labels** 10 pens 3 calculators	30 min for activity 20 min to discuss handout	3 students

*In this activity students will be moving between different locations, or stations, and a *station marker* is anything that can be used to mark the different stations. It is important that the station markers be heavy enough such that they don't blow away if doing the activity outside. Items that could serve as station markers include gym spots (small rubber mats), small marker flags, small cones, or pieces of paper.

**Mailing labels come in sheets. Cut the mailing labels into sheets of six; you need ten sheets of six labels each per team to play three rounds. You can also use sticky notes, but they often fall off of flying discs unless secured with tape.

Running the Activity

Before doing the activity, introduce your students to Euclid's algorithm, and make sure that your students are comfortable with calculating a quotient, q, and remainder, r. That is, if a and b are the larger and smaller integers respectively, then q and r are given by

$$q = \left\lfloor \frac{a}{b} \right\rfloor \qquad \text{and} \qquad r = a - bq.$$

Students will need to be able to do these calculations repeatedly when applying Euclid's algorithm.

Table 6.1: Euclidean Algorithm Examples

Round 1

Step	a	b	r	q
1	14304	3356	880	4
2	3356	880	716	3
3	880	716	164	1
4	716	164	60	4
5	164	60	44	2
6	60	44	16	1
7	44	16	12	2
8	16	12	4	1
9	12	4	0	3
10	4	0	n/a	n/a

Round 2

Step	a	b	r	q
1	35408	8213	2556	4
2	8213	2556	545	3
3	2556	545	376	4
4	545	376	169	1
5	376	169	38	2
6	169	38	17	4
7	38	17	4	2
8	17	4	1	4
9	4	1	0	4
10	1	0	n/a	n/a

Round 3

Step	a	b	r	q
1	26138	11094	3950	2
2	11094	3950	3194	2
3	3950	3194	756	1
4	3194	756	170	4
5	756	170	76	4
6	170	76	18	2
7	76	18	4	4
8	18	4	2	4
9	4	2	0	2
10	2	0	n/a	n/a

On the day of the activity, take the class to a location that has sufficient space to set up the stations as represented in Figure 6.2, leaving just enough room for students to toss the flying discs gently between stations. Have the students divide themselves into teams of three and designate the members of each team as A, B, and C. Each student should be carrying a calculator. Have each team set up their stations and get in their places for Round 1 as follows. Stations 1-10 each get marked with a station marker. Stations 1-9 each get a pen and a sheet of six mailing labels. Student A goes to Station 1 with two discs, and an extra sheet of mailing labels, using the mailing labels to mark one disc with the number 14304 (Step 1, a) and the other disc with the number 3356 (Step 1, b). Student B goes to Station 2 with the other disc and uses a mailing label to mark this disc with the number 3356 (Step 2, a). Student C goes to Station 3. This setup information is restated at the end of the activity as a checklist that could be photocopied and handed to students to expedite the setup process.

The students will work through Euclid's algorithm to find the greatest common divisor of two numbers over ten iterations. For the three rounds of play, the initial pairs of numbers, along with the interim solutions, are given in Table 6.1.

To begin Round 1, announce to the class that they are finding the greatest common divisor of 14304 and 3356, the two given numbers on their flying discs at Station 1. Then Student A divides the larger number on one disc by the smaller number on the other disc, finding the quotient, q, and the remainder, r. Student A writes the remainder, r, on two new mailing labels and uses these new labels to replace or cover the existing mailing label on each disc. He/she then tosses one of the discs to Student B at Station 2, tosses the

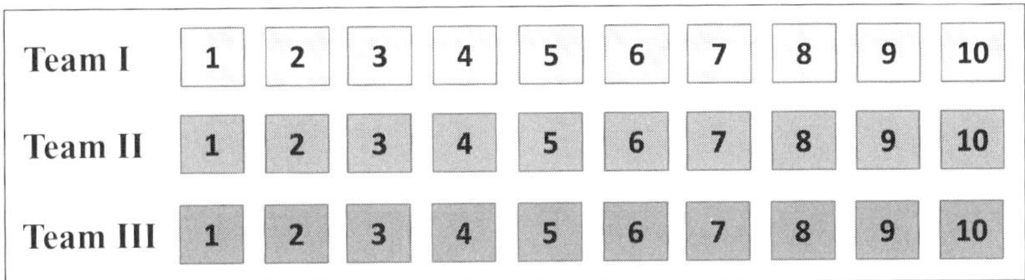

Figure 6.2: Layout for the station markers.

other to Student C at Station 3, and then runs to Station 4. Student B should have two discs; he/she divides the larger number by the smaller number, finding the quotient and remainder. Student B then writes the remainder on two new mailing labels, replacing the existing numbers with the newly found remainder. He/she then tosses one of the discs to Student C at Station 3, tosses the other disc to Student A at Station 4, and then goes to Station 5. Student C then divides the two numbers on his/her discs, replaces those numbers with the remainder, tosses one disc to each of the next two stations, and moves to the next open station. This algorithmic process (divide, replace, toss, and move) is repeated until both discs have reached Station 10, at which point the algorithm is complete since one of the discs should be labeled with a zero.

The activity is set up so each person has to do the steps in the algorithm three times in each round, and typically everyone figures it out by the end of the first round. The other rounds use the same setup, changing only the starting numbers.

At the end of the activity, have students fill out a master table like Table 6.1 for the last problem that they did. Then, working from the bottom row up, have them identify the greatest common divisor of each pair a_i and b_i until they are convinced that it is the same for each row. Once the physical portion of the activity is completed, provide students with the handout which builds on the concepts learned during the activity. This can be used for an in class discussion or can be completed as an out of class assignment.

Suggestions and Pitfalls

As students are doing the activity, you may want to check their remainders at each step. On the first round, you may also need to help weaker students.

To expedite the transition between rounds, you may want to fill out the labels for the initial flying discs ahead of time. The ideal class size is nine arranged into three teams of three. With larger classes, it helps to have a student assistant who can check answers for some of the teams. The station markers should be about 5-10 feet apart to allow room for tossing flying discs. Alternatively, a simplified version of this activity could be done in the classroom by passing notes instead of throwing flying discs.

This activity runs well by completing a total of three rounds – one round to make sure everyone understands the mechanics of the algorithm, and then, to bring out students' competitive spirits, the second and third rounds could be races. The examples provided in Table 6.1 use ten steps. This number could be increased or decreased to provide students with more or less practice.

Instructions for Initial Setup

In your team of three, each of you should select a role: Student A, Student B, or Student C. Then work together to complete the following steps to get ready for Round 1.

- Use station markers to designate ten stations that are 5-10 feet apart.
- Place a pen at each of the first nine stations.
- Place a sheet of 6 mailing labels at each of the first nine stations.
- Student A: Take a calculator, two flying discs, and an extra sheet of six mailing labels to Station 1. Once at Station 1, place a mailing label on each of the flying discs and write the number 14304 on one of the labels and 3356 on the other one.
- Student B: Take a calculator and one flying disc to Station 2. Once at Station 2, place one mailing label on the flying disc, and then use the pen to write the number 3356 on that mailing label.
- Student C: Take a calculator and go to Station 3.

Now you are ready to begin Round 1.

Finding the GCD – Class Handout

1. Run the algorithm with $a = 21,748$ and $b = 16,612$. Record a, b, r, and q at each step.

2. Run the algorithm with $a = 233$ and $b = 144$. Repeat for 144 and 89. Both pairs of numbers are adjacent Fibonacci numbers. Record a, b, r, and q at each step. What do you notice? Make a conjecture about what happens when you run the algorithm with any two adjacent Fibonacci numbers.

3. We define the sequences $\{a_i\}$, $\{b_i\}$, $\{q_i\}$ and $\{r_i\}$ to be the values for a, b, q, and r that the person doing the ith step has. Explain what the following three equations mean in terms of the activity.

$$a_2 = b_1 \qquad b_2 = r_1 \qquad r_2 = a_2 - b_2 q_2$$

4. Why does the activity stop when $b_n = 0$?

5. Suppose d is a positive integer. Prove that $d \mid a_1$ and $d \mid b_1$ if and only if $d \mid a_2$ and $d \mid b_2$. There are four proofs here. Two of them are short. Two of them are longer.

6. Why does the previous result tell us $\gcd(a_1, b_1) = \gcd(a_2, b_2)$?

7. Suppose when trying to find $\gcd(a, b)$ using this algorithm, $b_n = 0$ but $b_{n-1} \neq 0$. Justify each of these equations.

$$\gcd(a, b) = \gcd(a_1, b_1) = \gcd(a_2, b_2) = \cdots = \gcd(a_n, b_n) = \gcd(a_n, 0) = a_n$$

6.3 Symmetry and Group Theory with Plastic Triangles

Ron Taylor, Berry College

Concepts Taught: symmetry, group theory

Activity Overview

One way of describing introductory abstract algebra is to think of it as the theory that drives the high school algebra with which students are familiar. This foundational idea can be helpful when students begin to think about the process of solving equations that look like $ab^{-1}a = bxa^{-1}$. As students progress through the calculus sequence, they are often more interested in the results of their algebraic manipulations than in the process by which they are doing the algebraic manipulations. Once they begin studying abstract algebra, process returns to a place of central importance as they develop understanding of new structures based on the intuition they gain from performing calculations that look familiar.

This activity uses color-coded triangles to guide students through the development of a group table based on their intuitive understanding of rigid symmetries. This allows them to build upon what they know and to see some structure that has familiar aspects and, perhaps, unexpected ones. This activity can be done with the students in small groups or alone, with the freedom to discuss items with each other as they work.

Supplies Needed (per group)	Class Time Required	Group Size
A pair of different sized color-coded triangles*	30–50 minutes	1-3 students

*The smaller triangle should fit into, or onto, the larger one so that the student can manipulate each one separately or both of them at once. We have done this activity using triangles made of Zome®, as shown in Figure 6.3, but it can be done with triangles made of cardboard, or straws, or pipe cleaners, or anything else that is relatively rigid. The important thing is that there needs to be a way to distinguish between the vertices. Of slightly less importance is a way to distinguish the two sides, i.e., top and bottom, of the triangle. This ability to distinguish is one nice feature of the Zome triangles. We use colored nodes to distinguish the vertices and the shape of the Zome connectors to determine the top and bottom of the triangle. Figure 6.3a below shows the basic orientation of a triangle and Figure 6.3b shows the two nested triangles.

The pair of triangles allows for a notion of function composition as the symmetries are applied in turn to the inner triangle first and then to the pair of nested triangles. After the first symmetry is applied, the outside triangle shows the initial arrangement of the vertices that have been moved by the symmetry operation on the inside triangle. Then, once the

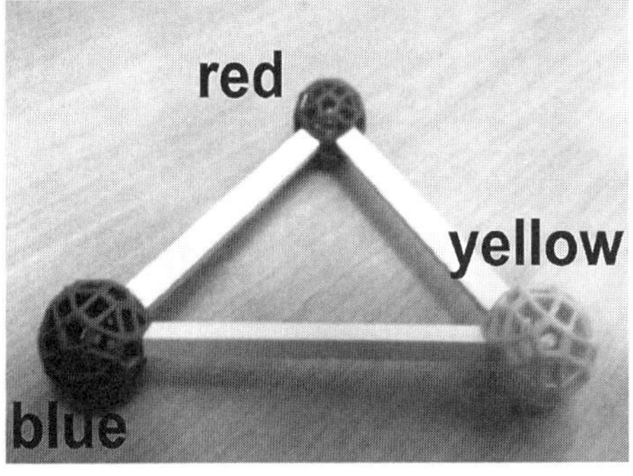

The top of a triangle.

Two nested triangles.

Figure 6.3: Triangles created with Zome.

second symmetry operation has been applied to both triangles at the same time, the outside triangle shows the second symmetry by itself and the inside triangle shows the composition of the two symmetries.

Running the Activity

After a brief discussion of the notion of symmetry as an action performed on an object, hand out the triangles. Students then work through the handout. The first problem is a simple question about solving a linear equation and explaining the solution without using the words *subtract* or *divide*. This is intended to steer the students toward talking about additive or multiplicative inverses.

Next the students are asked to think about all of the ways they can manipulate an equilateral triangle that maintain the initial orientation – a triangle with a bottom edge and a top vertex. They are able to use the arrangement of the colors of the vertices to distinguish between the different symmetries. The goal is to have them list the three rotations and the three reflections that preserve the orientation of the triangle. Meanwhile, they see that something like a $180°$ rotation would not be a symmetry because it would result in a triangle that has an edge on the top and a vertex on the bottom. Such orientations would be easily distinguishable from the original without having to consider the arrangement of the vertices. While looking for symmetries, the students are encouraged to describe each new result without worrying about notation so that intuition-building is their main focus.

After students have spent some time thinking about ways of manipulating the triangles, Problem 5 asks them to consider how each of the six symmetries would rearrange the colors of the vertices. They record their observations on the diagram as a guide to complete the group table. Then in Problem 6, they are asked to put the smaller triangle inside the larger triangle so that both of them are in the standard position with red at the top, yellow at the bottom right, and blue at the bottom left. In order to complete the group table they are instructed to think of the symmetry listed along the left column of the group table as being applied to the inner triangle by itself and then the symmetry along the top row being applied to both triangles at the same time.

After the group table is complete, the students are asked to make some observations. Since the top row and left column match the header row and column of the completed table, they can see that R_0 is the identity element. In addition, because R_0 appears in each row and column, they can deduce that each symmetry has an inverse. The fact that the table itself is not symmetric across the diagonal points to the fact that the composition of symmetries is not commutative. In the final table there are blocks of Rs and Fs. This can be a foundation for discussion of subgroups and cosets later in the course and gives a nice geometric foundation for this idea based on how rotations and flips interact with themselves and each other. At the end of the handout, the students are asked to solve an equation similar to the initial simple example. In this case they do not have access to the words *subtract* or *divide* since they have not been defined, so they are forced to consider multiplying by an inverse element. This reinforces the notion that the operation is not commutative and forces students to think about where the multiplication should take place on each side of the equation.

Suggestions and Pitfalls

As students go through the activity, they need to be mindful to always return both triangles to the initial configuration before attempting to perform the symmetry actions in sequence. This is particularly important as they record their answers in Problem 5 for the possible arrangements of the colors in the six individual symmetries.

Symmetry and Group Theory with Plastic Triangles – Class Handout

Definition. A *symmetry* is an action which brings every point of an object to a position previously occupied either by itself or by some other point of the object.

1. Explain how you would solve the equation $5x + 8 = 21$ without using the words *subtract* or *divide*.

The following items deal with the symmetries of an equilateral triangle. The triangle shown to the right has colored vertices. These colors are there to help you keep track of what is going on as you complete the exploration that follows. Throughout the rest of the activity, please think of this configuration as the starting point.

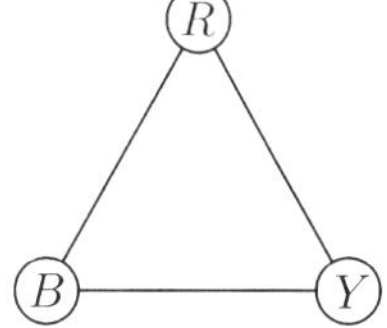

2. How many symmetries does an equilateral triangle have? That is, how many ways can you move the triangle around to get a triangle that matches the orientation of the starting triangle? (This means you want to start and end with a triangle that has a bottom edge and a top vertex.) Don't worry about notation, just use words to explain what you mean.
3. Can you think of any similar actions on an equilateral triangle that are not symmetries?
4. What happens if you perform one symmetry on an equilateral triangle followed by another one?
5. Using one of the triangles you have, determine and annotate the new locations of the colors after you apply the symmetry given in each picture below; record your answer on each picture by marking each vertex with R, Y, and B. Start with your triangle oriented as indicated in the image above and fill in the circles with the appropriate colors as they appear after you apply each symmetry.

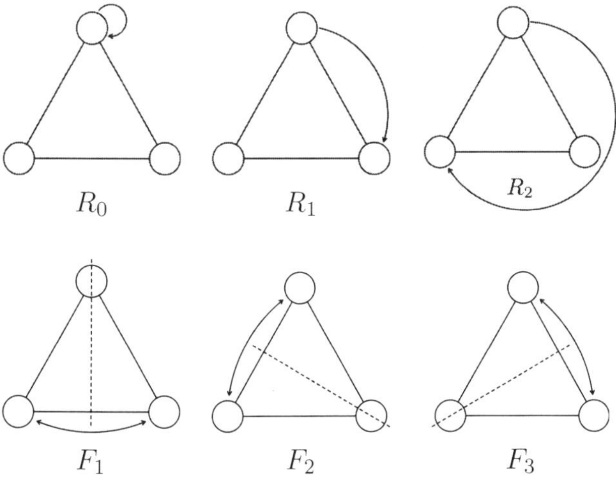

6. Fill in the table below by performing the symmetry along the left column first, followed by the symmetry along the top row. As you begin, put the smaller triangle inside the larger one and then, as you carry out the symmetry actions, perform the first symmetry on the inside triangle only and then perform the second symmetry on both triangles at the same time. Record what happens to the inner triangle. We have filled in the box corresponding to $R_2 \bigotimes F_2$. (Each entry in the table should be one of the following: $R_0, R_1, R_2, F_1, F_2, F_3$.)

$\bigotimes$	R_0	R_1	R_2	F_1	F_2	F_3
R_0						
R_1						
R_2					F_1	
F_1						
F_2						
F_3						

7. Is there an identity symmetry? If so, what is it? Does each symmetry have an inverse?

8. Is the operation $\bigotimes$ commutative? If so, how can you tell? If not, why not?

9. Are there any interesting patterns in the table? What do these patterns tell you about the operation $\bigotimes$?

10. Explain how you would solve the equation $F_2 \bigotimes x = R_2$ using the ideas you developed in this activity. How would you represent the solution to this equation symbolically?

6.4 Finding Groups in a (New) Color Cube Puzzle

Justin Brown, Olivet Nazarene University
Dale Hathaway, Olivet Nazarene University

Concepts Taught: product groups, group generators, relations

Activity Overview

Most students in an abstract algebra course appreciate any activity that provides tangible insight into the theoretical nature of the material. While there are many games and puzzles that can be used as an illustration of groups, the following color cube activity allows students to discover a group that models the scenario; the groups found are small enough that the students can work out all of the details by hand. Surprisingly, the group that a student finds, and the relations on the elements of that group, may vary based on the original setup that the student chooses. Students are often intrigued by the diversity of correct answers discovered by themselves and their peers.

In [1], Snape and Scott pose the following puzzle, "Take 27 cubes, three each of nine different colors, and construct a $3 \times 3 \times 3$ cube such that each color is represented exactly once on each face." After solving a simplified version of this puzzle and the puzzle itself, students permute the layers of the $3 \times 3 \times 3$ cube to lead to other solutions of the puzzle. Thus, one can view these various permutations on the cube as a group.

Supplies Needed	Class Time Required	Group Size
Container of color cubes*	20-40 minutes	1-2 students

*Each student or pair of students receives 27 cubes, three of each of nine different colors. Containers of color cubes may be purchased from any learning-manipulatives seller.

Running the Activity

Provide students with 27 color cubes and a copy of the handout which is designed to guide the students through the activity. Part 1 of the handout has the students choose four colors, set aside two cubes of each chosen color, and create a $2 \times 2 \times 2$ cube such that each color appears exactly once on each face (see Figure 6.4). Once they have created such a cube, they are asked to describe their setup by writing the color of each block in a table. This setup is called their *original state*. Students are told that their original state must not match anyone else's in the class.

The elements of the groups that students examine are operations on the cube that lead to different arrangements of the colors that still satisfy the condition of each color appearing exactly once on each face. The group operation is composition of the operations

Figure 6.4: Example of a $2 \times 2 \times 2$ cube and a $3 \times 3 \times 3$ cube.

	Id	L	F	$L \circ F$
Id	Id	L	F	$L \circ F$
L	L	Id	$L \circ F$	F
F	F	$L \circ F$	Id	L
$L \circ F$	$L \circ F$	F	L	Id

Figure 6.5: Cayley table for the $2 \times 2 \times 2$ case.

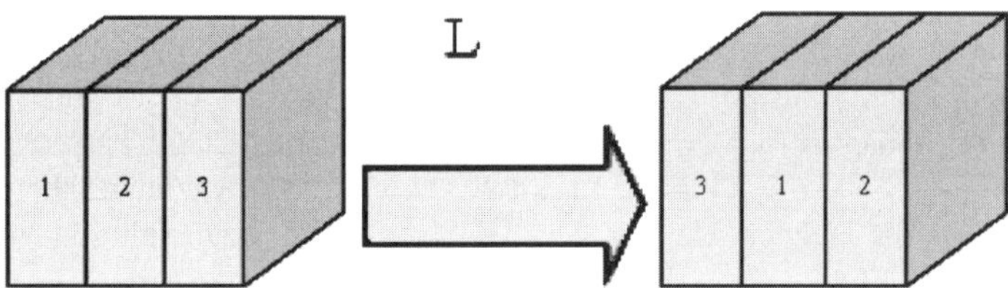

Figure 6.6: One way to visualize "L."

on the cube. As students complete the rest of Part 1 of the handout, they find the color arrangements corresponding to group elements L and F (as defined in the handout) and any compositions of L and F. In Problem 2c they are asked to compute the Cayley table for their group. They should obtain the table shown in Figure 6.5 which is isomorphic to $\mathbb{Z}_2 \times \mathbb{Z}_2$, the Klein four-group. Lastly, the handout defines a new operation B, and students discover that $B = L \circ F$; thus it is not necessary to include B as a generator of their group.

Once students are comfortable with the $2 \times 2 \times 2$ case, have them begin Part 2 of the handout in which they are asked to create a $3 \times 3 \times 3$ cube using all nine colors. As in Part 1, students will record their original state and find the color arrangements corresponding to L, F, and B as defined in the handout, and all possible compositions. In case there is any confusion about how the operations from the $2 \times 2 \times 2$ case translate to the $3 \times 3 \times 3$ case, show the class Figure 6.6 representing the action associated to L or demonstrate the action for them. Unlike the $2 \times 2 \times 2$ exercise, different original states can lead to different relations between B, L, and F, and sometimes to no relation at all! Since some students will obtain 27 element groups while others will have 9 element groups, students are not asked for a Cayley table for the $3 \times 3 \times 3$ problem.

Suggestions and Pitfalls

The process of actually writing out the $3 \times 3 \times 3$ group can be tedious; we recommend cutting down the class time required by having the students do this as homework.

As already mentioned, the instructor should be aware of the fact that different original states often lead to different solutions to the last problem. When checking to see if a student has done the assignment correctly, the instructor may be able to notice certain patterns in the original state that indicate whether a student should find a relation between B, F, and L, or whether no relation exists.

A simpler version of this activity involves only two colors: three cubes of one color (say white) and 24 cubes of a second color. Students then set up a $3 \times 3 \times 3$ cube such that each face contains exactly one white cube. With this setup, we expect that all students will obtain the same group, though relations may still vary.

Of course, there are other operations we have not mentioned that preserve the puzzle, which if introduced will make the obtained groups larger. For instance, one may introduce permutations of the slices of the cube, in which case the groups discovered will contain subgroups isomorphic to S_3 instead of $\mathbb{Z}_3$. One could also rotate the entire cube as a group element.

When the activity is finished, many students are curious about the $4 \times 4 \times 4$ case. However, working with this size cube requires stricter definitions on the original state since there are eight hidden interior cubes.

Reference

1. C. Snape and H. Scott, *Puzzles, Mazes and Numbers*. Cambridge: Cambridge University Press, 1995.

Finding Groups in a (New) Color Cube Puzzle – Class Handout

Part 1: $2 \times 2 \times 2$ Cubes

1. Choose four of your colors, and gather two cubes of each of these colors. Create a $2 \times 2 \times 2$ cube such that each color appears exactly once on each face. Describe your setup (called the *original state*) by writing the color of each block in a table like the one that follows. Your original state should be unique to you; do not use the same color setup as anyone else!

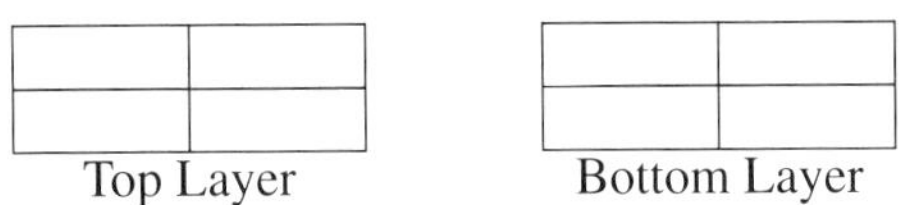

2. The elements of the groups examined will be operations on the cube that lead to different arrangements of the colors that still satisfy the condition of each color appearing exactly once on each face. The group operation will be composition of the operations on the cube. The cube has six faces; we will consider each of these faces to be a "slice." Orient the cube so that there is a left slice, a right slice, a front slice, a back slice, a top slice, and a bottom slice.

   (a) Take the slice of blocks on the right and place it on the left maintaining the slice in its original orientation. Does this give a new arrangement of colors? This group element is labeled L.

   (b) Return the cube to its original state. Similarly, take the slice of blocks in the back and move it to the front. Does this give a new arrangement of colors? This group element is labeled F.

   (c) Now consider all color arrangements that are possible to obtain by composing L and F. Is this a group? If so, to which known group is it isomorphic? Give the Cayley table.

   (d) Return the cube to its original state. Move the top slice of blocks to the bottom and call this group element B. Is B truly a new group element, or have you obtained a color arrangement that we already saw in part (c)? If so, write the equation for B in terms of L and F.

Part 2: $3 \times 3 \times 3$ Cubes

1. Now use all 27 cubes. Create a $3 \times 3 \times 3$ such that each color appears exactly once on each face. Describe your setup (called the *original state*) by writing the color of each block in a table like the one that follows. Your original state should be unique to you; do not use the same color setup as anyone else!

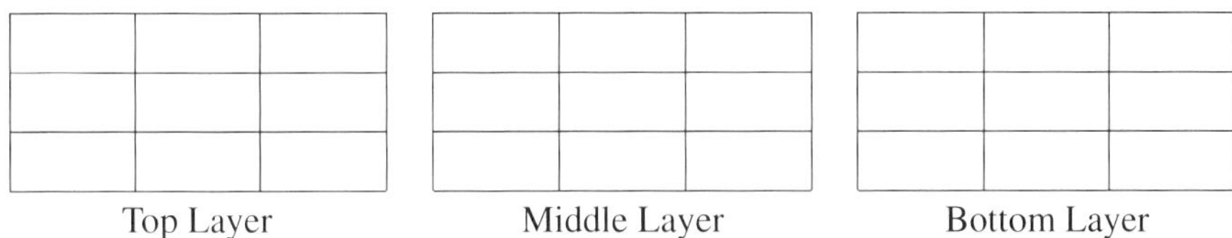

Top Layer Middle Layer Bottom Layer

2. We will define elements of a group the same way as in Part 1. L is taking the slice on the right and placing it on the left. F is taking the slice on the back and placing it in the front. Now consider all of the color arrangements that are possible to obtain by composing L, F, and powers of L and F. To which known group is this isomorphic?

3. Return the cube to its original state. Define B as taking the top slice and moving it to the bottom. Is B already a group element in your answer to the previous problem? If so, write the equation for B in terms of L and F. If not, what new group do you obtain by looking at all color arrangements that are possible by composing L, F, B, and their powers? *Answers will vary based on your original state; therefore, your results may differ from your classmates.*

6.5 Acting Permutations

Julie Beier, Earlham College

Concepts Taught: permutations

Activity Overview

While students may quickly grasp how to compute with permutations, they often struggle with the move from a concrete computation to abstract concepts including the use of notation and the difference between the permutation as an element and the elements permuted. In this activity, students develop intuition for the way permutations behave by acting them out in a guided setting that allows them to uncover and explore many of the general properties of permutations.

Supplies Needed	Class Time Required	Group Size
Index cards Number signs* Masking tape or painter's tape Flip chart and marker	50-75 minutes	Whole class

*Before the activity, make *number signs* with plain paper. A number sign is a piece of paper with a number on it. Use a marker to write one number on each piece of paper using the numbers 1-6 each twice. For larger classes, you may choose to make number signs with 1-7 or 1-8 instead.

Running the Activity

Before doing the activity, introduce permutations using two-line notation, such as

$$\begin{pmatrix} 1 & 2 & 3 & 4 & 5 & 6 \\ 3 & 2 & 6 & 4 & 5 & 1 \end{pmatrix}.$$

It is helpful to present several illustrations such as an action of the dihedral group or a problem assigning chores to people. Ask students to think about different permutations of six elements. These permutations will be used for the activity. Invite students to share their permutations in groups of three to ensure that the permutations contain a diverse collection of interesting features such as fixed elements, elements that are part of smaller cycles, permutations that decompose, or permutations that use every number. Have students record their permutations on index cards (one per card). Collect the cards and go outside or to a location with room to move around, like a gym or large hallway.

Use tape to secure a set of number signs on the ground in a slightly curved arc. Ensure that the numbers are distanced so that students will have reason to run if they are moving from 1 to 6. Also, set up a flip chart or large portable dry-erase board to share permutations and record findings. This should be placed such that all students in the line of numbers can see it.

Give the students the handout which has prompts for exploring ideas as they emerge during the activity. Assign 6-8 students to be runners for the first portion. Give each runner a number, and have them stand at the corresponding number on the ground. The runners are responsible for being permuted and thinking about how they know what to do. Students who are not runners are observers; their job is to check that the runners move correctly, look for patterns across examples, and report their observations. If you believe that your students may struggle to permute correctly, assign an observer to each runner. Explain the roles to students.

Write a permutation in two-line notation and have each student run on the count of three to the number listed in the notation just below the number he/she is holding. One – two – three – permute! Then, ask students how they know that they are in the right spot. It may take several tries to get the first permutation correct, and students may confuse label with position. Either have students reset and return to their initial positions, or select new runners; then act out a different permutation from the cards of student-created permutations. Repeat. As interesting features emerge, have students take note of their observations.

At this point, there are many ideas that can be explored. Some easy options are: other notation conventions, the fact that permutations are non-commuting in general, the identity permutation, inverse permutations, order, the fact that permutation expression is non-unique, transpositions, and the parity of permutations. Ask students to refer to the handout and read the desired prompt; facilitate a discussion about it as students act out examples. Allow students time to record their conjectures and answers during the activity. The prompts may be introduced in any order except for the case of 2-cycles, which should be last. While you may switch runners at any time, it is convenient to switch runners when exploring a new topic.

This physical realization of permutations empowers students to generate their own conjectures. As students pose conjectures, immediately test them, as seen in Figure 6.7. This allows the class to quickly refine their ideas and prevents attachment to incorrect concepts. While doing the activity, record conjectures and note counterexamples as they are found. After completing the activity, encourage students to try to prove any of the conjectures that have not been disproven. These examples and ideas will be useful touchstones for reference as you continue to discuss permutations, the symmetric group, and other finite groups.

Suggestions and Pitfalls

If you have fewer than six students, the activity can be completed with colored candy. For example, use M&M's and large paper with locations numbered either 1-6 or by color. The same procedures can be acted out and the same ideas discovered.

Avoid immediately correcting student errors. Push the student observers to locate these errors. Similarly, allow students to make conjectures that will prove to be false and lead them to counterexamples. You may need to remind students to stay on task. While many of the topics can be covered in one hour, it takes an extra 20-30 minutes for students to discover cycle parity on their own. However, this discovery is typically powerful for them and helpful in understanding later theorems that often feel "magical".

Photos courtesy of Susanna Tanner, Earlham College.

Figure 6.7: Students generating a conjecture and then testing it.

Acting Permutation – Class Handout

Below are some things to consider as we explore permutations. Record your conjectures and observations. For homework, try to prove at least one of these conjectures.

Notations for Permutations: In two-line permutation notation, the top line always stays the same. Are there more efficient ways to record a permutation? What if instead of viewing a permutation as recording where all of the things move, we think of a permutation as a chain of events? For example, can we record what happens to the element 1 over time? How would we do this?

Commuting: Suppose we want to compose permutations like we do functions. What would it mean to do this and what would we get? Record two permutations from class and their composition. Does order matter in the composition of permutations? Provide an example that illustrates your answer. Will permutations ever commute? What conditions would be necessary for them to commute?

Inverses: What would it mean to "undo" a permutation? What does that mean in terms of the students who are acting out the permutation? Suppose I give you a random permutation. How can you find its inverse? Will your method always work?

Identity: Is there an identity permutation? What is it? How would you write it down? Write it down using all of the different notations we have discussed.

Order: Write down a permutation that consists of one cycle and act it out. What is the order of this cycle? In other words, how many times do you need to apply this permutation to get everyone back to their starting positions? Is there a general rule that will tell you the order of any permutation? Try your rule on permutations consisting of one cycle and many cycles.

Lack of Uniqueness: Record a permutation that the class acted out. Is this the only way to record what happened? If not, how many ways can you write it down? Can you develop a way that uses more or less cycles?

2-Cycles and Cycle Parity: A 2-cycle, or transposition, is a permutation that only switches two numbers. What is the inverse of a 2-cycle? Since we know that cycle notation is not unique, can you write the identity as a product of transpositions? Find three ways to do this. Can you write the identity using an odd number of 2-cycles? Why not?

More 2-Cycles: Consider the permutation (351426). Can you write this permutation using only 2-cycles? What is the fewest number of transpositions you can use? It may help to act this out. Is there a way to write this permutation with an even number of 2-cycles? What about an odd number? Why or why not?

Challenge: Find an algorithm for writing a permutation as the product of transpositions. Will a permutation only give you an odd or even number of transpositions? Why or why not?

6.6 Nametags and Derangements: A Class Permutation

Ann N. Trenk, Wellesley College

Concepts Taught: derangements, Stirling numbers, permutations

Activity Overview

In this activity, we use sticky notes and students' hands to create a permutation of the set of students in the class. The arrangement of the students demonstrates the cycle notation for this permutation. In combinatorics, this activity can be a starting point for studying derangements or Stirling numbers. It is also a useful way to engage students in thinking about asking good questions in mathematics and modeling what mathematicians do as they think of research questions.

Supplies Needed	Class Time Required	Group Size
A sticky note for each student	10-15 minutes	≤ 30 students

Running the Activity

Let n be the number of students participating in the activity. Number n sticky notes with the numbers 1 through n, and have each student write his/her name on one of them so it becomes a nametag (Figure 6.8a). Turn the nametags face down and mix them so that the names are hidden. Alternatively, fold them in half and put them in a hat or bowl. Have each student take one sticky note and give them these instructions:

"With your right hand, seek out the left hand of the person whose nametag you picked." It is helpful to add, "Keep in mind that as you are doing this, someone else is doing what?" They can answer, "seeking out your left hand."

A mad scramble ensues as they look for one another and try to remember the roles of the right and left hands. Some students don't know each other's names and the instructor can help them locate one another. Often a student will ask what they should do if they get their own nametag, and you can reply that they should be able to follow the instructions since they have both a left hand and a right hand.

Indicate the configuration of circles formed by listing students' names using their numbers as shown in Figure 6.8b, or just recording the sizes of the resulting circles. An example with nine students is shown in Figures 6.8 and 6.9. In this example, two students haven chosen their own nametags.

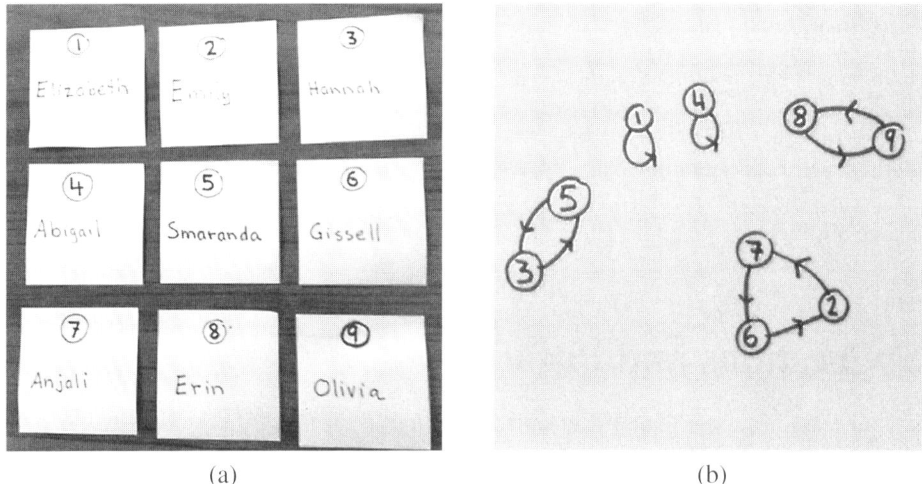

(a) (b)

Figure 6.8: (a) The prepared nametags. (b) One possible permutation.

Figure 6.9: The nine students acting out the permutation in Figure 6.8b.

After the activity, students return to their seats. Lead a class discussion about the value of asking good questions in mathematics and how this can lead to interesting mathematical research. Give them a few minutes to write down some counting questions about the nametag activity and use these to lead into a discussion of derangements and later Stirling numbers. The following are some good questions studen ts in a combinatorics class might ask, perhaps with some prompting. Note that the last three of these questions cannot be answered quickly and could lead into deeper topics to be explored later.

- How many possible outcomes are there for this activity? ($n!$)
- How many of these outcomes result in one big circle? ($(n-1)!$)
- In how many outcomes does at least one person get his/her own nametag?

- In how many outcomes does nobody get his/her own nametag?
- How many outcomes result in exactly k circles?

Once the activity is completed, students further explore the concepts learned by completing one of the handouts. Page 257 contains two half-page handouts, each with a different focus. These are included because the activity fits well both in a combinatorics class and an abstract algebra class. These handouts could be used as homework assignment or as an in class group assignment.

Suggestions and Pitfalls

If your class has more than 30 students, you might want to try this activity with smaller groups rather than with the whole class.

You will need some open space for this activity. If your classroom does not have enough open space, you can move to a hallway or even outside.

Sometimes the groups get tangled, and need to be separated so that each circle of students has its own area of the room. Occasionally, a circle of students looks like a figure eight and needs to untangle itself so that it looks like a circle.

Nametag Combinatorics Activity – Class Handout

We just performed the following activity in class. Each student wrote his/her name on a sticky note to create a nametag. The nametags were mixed and each student chose one. Then, with his/her right hand, each student sought out the left hand of the student whose nametag he or she picked. The result was one or more circles of students.

1. How many possible outcomes are there for the nametag activity with n students?

2. How many of these outcomes result in one big circle?

3. When the nametag activity is done with 20 students, what is the probability that the outcome is one big circle? Simplify your answer, but don't approximate.

4. When the nametag activity is done with 20 students, how many outcomes have one circle of 12 students and one circle of 8 students? What is the probability of this occurring? Simplify your answer, but don't approximate.

Nametag Abstract Algebra Activity – Class Handout

We just performed the following activity in class. Each student wrote his/her name on a sticky note to create a nametag. The nametags were mixed and each student chose one. Then, with his/her right hand, each student sought out the left hand of the student whose nametag he or she picked. The result was one or more circles of students.

Now you can try a smaller version of the nametag problem individually or in a small group. Each group should have small squares numbered $1, 2, \ldots, 10$.

1. Turn your squares face down so you cannot see the numbers. Mix them. Turn over the squares one at a time and record each number you get in the chart below. The first number you turn over goes under the 1, the second number goes under the 2, etc.

1	2	3	4	5	6	7	8	9	10

2. Draw the circles that result from your chart. You can think of your drawing as a top down view of the nametag activity.

6.7 Discovering Catalan Numbers Using M&M's®

Ann N. Trenk, Wellesley College

Concepts Taught: Catalan numbers, combinatorics

Activity Overview

The Catalan numbers ($c_0 = 1$, $c_1 = 1$, $c_2 = 2$, $c_3 = 5$, $c_4 = 14$, $c_5 = 42$, etc.) are a fascinating sequence of numbers that arise in different contexts, and as a consequence have been rediscovered many times. One way to define c_n is the number of sequences of n red and n green M&M's so that for each k, among the first k M&M's in the sequence, there are never more greens than reds. This is a complicated counting question and students in combinatorics often have an easier time understanding such a question if they first list some sample elements in the set they are trying to count as well as some elements that fall outside this set. This activity provides such an opportunity.

There is a lovely recursive formula for the Catalan numbers ($c_n = c_0c_{n-1} + c_1c_{n-2} + \cdots + c_{n-2}c_1 + c_{n-1}c_0$) as well as an explicit formula $c_n = \frac{1}{n+1}\binom{2n}{n}$. The handout following this activity leads students to conjecture the explicit formula. The activity also can be helpful when presenting a proof of the recursive formula.

Supplies Needed	Class Time Required	Group Size
Red and green M&M's 1 container per student* Yarn**	10-15 minutes	Whole class

*Cupcake liners work well, but anything small that can hold six M&M's, such as a plastic bag or a bowl, is fine.
**The yarn needs to be long enough so that when the students are standing shoulder to shoulder, there is yarn in front of all of them. This yarn marks the edge of a mud pit. You can have one student hold the end of the yarn and another unravel it so it is stretched out in front of the whole class.

Running the Activity

For each student, prepare a container with three red and three green M&M's as seen in Figure 6.10. Have the students line up, shoulder to shoulder, holding their M&M's and facing an imaginary mud pit (Figure 6.11a). Ask the students to close their eyes and choose a piece of candy from their bowl. Once they have chosen the candy, they can open their eyes, look at its color, and eat or discard it. Have those students who chose a red candy take

Figure 6.10: M&M's placed in cupcake liners before starting the activity.

a step backwards away from the mud pit; have those students who chose a green candy take a step forward into the mud pit and kneel down to symbolize that they are stuck in the mud. Once a student is in the mud pit, he or she is done taking steps, but can continue to pick and eat M&M's with the rest of the class. The situation at this point is illustrated in Figure 6.11b.

The activity continues in the same way. Again the students close their eyes, pick a candy, then open their eyes and look at its color. If it is green they take a step forward, and if it is red they take a step back. All steps are the same size. After eating two pieces of candy, students who chose GG or GR will be kneeling down in the mud pit, students who chose RG will be back at the edge and those who chose RR will be two steps back from the edge. The activity ends when the students have picked all six of the M&M's. Take note of how many are kneeling down and how many are safe. Students should observe that those standing are back in their starting position at the edge of the mud pit.

After the activity, the students are prepared to work through the handout which guides them to conjecture the explicit formula $c_n = \frac{1}{n+1}\binom{2n}{n}$. The activity is also helpful if the instructor chooses to give a combinatorial proof of the recursive formula $c_n = c_0c_{n-1} + c_1c_{n-2} + \cdots + c_{n-2}c_1 + c_{n-1}c_0$. In that proof, the term $c_{k-1}c_{n-k}$ counts the number of safe sequences in which the first return to the edge of the mud pit is after eating exactly k red and k green candies, and this can be described by placing a safety line one step back from the edge of the mud pit.

Suggestions and Pitfalls

If you have a large class, or a small classroom, you can do the activity twice with half the class participating each time and the other half observing. If the weather is nice, you could bring the class outside and use the edge of a walkway as the edge of the mud pit.

(a)

(b)

Figure 6.11: (a) Students at the beginning of the activity. (b) After eating some M&M's.

M&M's® Activity – Class Handout

Clifford stands at the edge of a mud pit, facing the edge and holding a bag containing n red and n green M&M's. He draws out the candies one at a time and eats them. If he draws a red one, he takes a step back. If he draws a green one, he takes a step forward. All steps have the same size.

1. How many different arrangements of n red and n green M&M's are there?

Call a sequence of n red and n green M&M's *safe* if it leads to Clifford remaining at the edge of the mud pit. For example, when $n = 4$, the sequence $RRRGGGRG$ is safe, but $RGGRRRGG$ is not.

2. For each of the specific cases $n = 1, 2, 3$, list all safe sequences.

3. For each of the specific cases $n = 1, 2, 3$, calculate the probability that Clifford does **not** go over the edge. *Note that the probability a sequence is safe is* $\frac{\#\ \textit{safe sequences}}{\textit{total}\ \#\ \textit{sequences}}$.

4. Based on these cases, develop a conjecture for the probability that he stays safe in the general case.

5. Use your conjecture from (4) and your answer to (1) to develop a conjecture for the number of safe sequences in the general case.

6.8 Walking the Seven Bridges of Königsberg

Hannah Robbins, Roanoke College

Concepts Taught: graph theory, modeling

Activity Overview

In this activity students explore Euler's classic Seven Bridges of Königsberg problem by walking around on a map of the city. The activity serves as a great introduction to graph theory by creating a mathematical model of a practical situation and asking students to explore the idea of proving that something cannot be done.

Supplies Needed	Class Time Required	Group Size
2 colors of duct or masking tape* (several rolls per color)	45 minutes	2-3 students

*Before class, use the tape to make copies of the map on the floor as shown in Figure 6.12. Make one copy of the map for each group. Be sure that the maps are large enough so that it is easy to walk around on them (Figure 6.12b); the bridges seen in Figure 6.12b are about 20″ long and 8″ wide. One color of tape, preferably blue, is used for the river and the second color for the bridges. Number the bridges, making sure your bridge numbering is consistent between all maps and matches the class handout. (Making all the maps will take longer than you think, so be sure to leave enough setup time. The map shown in Figure 6.12b takes 25-30 minutes to make, although less elaborate maps with a narrow bridge or narrow water would take less time.) Alternatively, you could use sidewalk chalk and do this outside with less setup time.

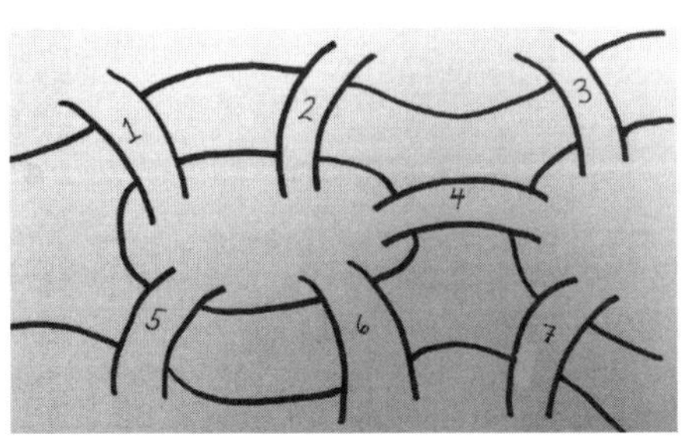

(a) Picture of map.

(b) Picture of tape version of map on floor.

Figure 6.12: Setting up the problem.

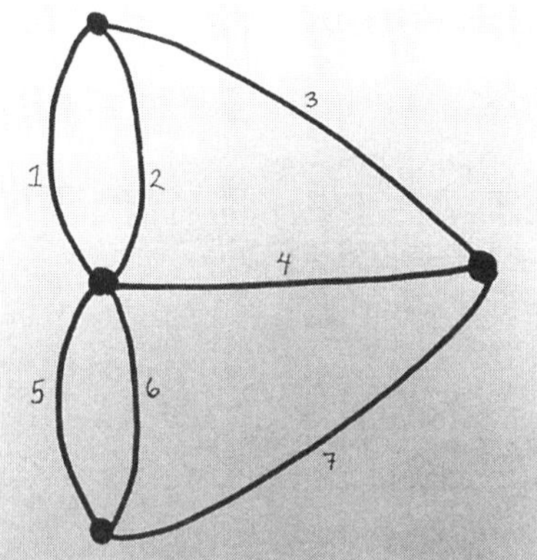

Figure 6.13: Modeling the problem with a graph.

Running the Activity

Introduce the activity by showing students the maps on the floor. Ask the class: Is it possible to walk through this city in such a way that you cross each bridge exactly once? Is it possible to do this and end up back where you started? The only rules are: (1) you can only cross the rivers by using a bridge and (2) once you start to cross a bridge you must cross it completely, no going halfway across, turning around and going back.

Pass out the handout which describes the activity, divide the class into groups of 2-3 students, and assign each group to one of the maps. Let the groups try to find a path crossing every bridge once. You may want to suggest that one group member walk on the map and another keeps track of which bridges they've crossed. Point out that their path can start anywhere on the map. Let the groups work until they are feeling fairly frustrated and are convinced that the problem cannot be solved.

Reconvene as a class and help them model the map using a graph whose vertices are the separate land masses (top, bottom, right, and island) and whose edges are the seven bridges as seen in Figure 6.13. Have the students work towards translating the original questions into finding sequences of edges in this graph where each edge is used exactly once. Introduce the idea that edges in the middle of such a path must occur as "in-out" pairs, i.e., whenever your path travels to a vertex it must come in along one edge and go out along another. The first and last edges do not need partners. Then, guide the class to the conclusion that each vertex in the middle of our path connects to an even number of edges. (If the path starts and ends at the same vertex then all vertices have an even number of edges.) Discuss why the graph or map has too many vertices with an odd number of edges, and how this proves there are no such paths in a way that simply failing to find them did not. Ask the groups to use their new criteria for when such paths exist to fix our map both by adding a new bridge and by destroying an existing bridge.

Suggestions and Pitfalls

This activity is written using terminology that is accessible for any level course. If you are using this activity as an introduction to a more detailed examination of graph theory, you can introduce technical vocabulary here (vertex, Euler path, etc.); this also makes a nice lead-in to either Fleury's or Hierholzer's algorithm.

For more information on the classic Seven Bridges of Königsberg Problem, see Chapter 1 of [1].

Reference

1. N. L. Biggs, E. K. Lloyd, and R. J. Wilson, *Graph Theory 1736-1936*. Oxford University Press (Clarendon Press), 1999.

Walking the Seven Bridges of Königsberg – Class Handout

On the floor is a map of a city with seven bridges. Can you find a path through this city which crosses each bridge exactly once? Can you do this and end up back where you started?

The only rules are simple: You can only cross the rivers by using a bridge (no swimming!). Once you start to cross a bridge, you must cross it completely (no going halfway across the bridge, turning around and going back).

Below are several copies of the map with bridge numbers so that you can keep track of paths as you try them. You may want to have someone walking the map and someone else records their path. As you work, consider the following questions:

1. What happens if you start on different parts of the map? You can start your path on either bank of the river, the island in the middle, or the peninsula on the right.

2. Are there symmetries of the map that can help you reduce the number of different paths to check?

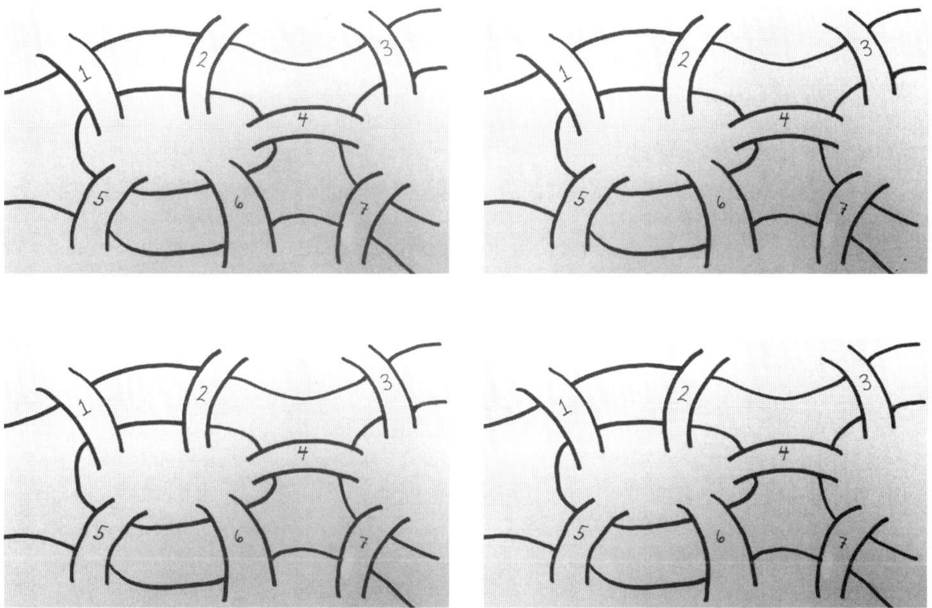

6.9 Designing Round-Robin Tournaments Using Yarn

Jill Bigley Dunham, Chapman University

Concepts Taught: graph theory, edge coloring, tournament design, round-robin tournaments, algorithmic thinking

Activity Overview

This activity provides a colorful introduction to graph theory, a topic that is unfamiliar to many undergraduate students. As students explore the design of round-robin tournaments, they have the chance to work together in groups on a problem that is often of interest to them. There is a lot of information to track when designing a tournament, and the problem becomes complicated quickly. In particular, when trying to extend a four-person tournament to include six people, students typically are challenged to find a solution; this creates an opportunity to understand the complexity and the need for deeper mathematical tools. To see the situation more clearly, students organize themselves into a round-robin tournament using colored yarn to specify rounds of the tournament. As the problem becomes larger, an algorithm is introduced.

This activity provides students with an opportunity to develop an intuitive understanding of the basic ideas of graph theory and how they can be used to solve problems. The terminology of graph theory can be introduced during this activity, or you can work through it without ever using any technical language.

Supplies Needed (per group)	Class Time Required	Group Size
42 strands of colored yarn* 7 colored pens* 1 gallon-sized sealable bag*	50 minutes	8-10 students

*Each group will need six two-meter long strands of each of seven colors. They also need seven different colored pens to match the seven colors of yarn. Distribution of supplies runs most smoothly if all the materials for each group are sealed into a gallon size plastic bag prior to class.

Running the Activity

The handout for this activity comes in two parts. Part I guides students through a whole-class discussion and an introductory small group activity. Part II describes an algorithm to use for larger tournaments. To keep from giving away the algorithm too early, only distribute Part I at the beginning of class.

Figure 6.14: Students creating a tournament graph.

Begin the activity by doing a 5-10 minute demonstration with four volunteers labeled as A, B, C, and D. Explain that a round-robin tournament is one in which each participant plays each other participant exactly once. Then ask the class to describe how they could do a round-robin tournament among the four volunteers. Tell your students that to visualize the problem, we can use a piece of colored yarn to represent a game. Two students holding the ends of the yarn are playing against each other; the color of the yarn indicates the round. Let red yarn stand for the first round.

Give students A and B each an end of a piece of red yarn to indicate that they will play a game against one another in the first round. Similarly, have students C and D hold another piece of red yarn indicating that they are also playing a game against one another in the first round. This obeys the rules of the round-robin tournament. Talk the students through the design of the four-person tournament, as they continue to build the visual model by adding a new color of yarn for each round. Ask students how many rounds are needed? How many total games are played? Figure 6.14 shows students using yarn to represent games in a tournament.

Next, solicit two more volunteers, label them E and F, and ask your students to design a six-person tournament. An obvious first move here is to give students E and F a piece of red yarn, representing a game in Round 1. Let them try to do a few more steps and see what goes wrong.

Have students form groups of 8-10 people, and provide each group with a bag of pens and yarn. Each group should first complete Problem 1 from the handout, recording their observations related to the classroom demonstration and using the colored pens to keep track of the colors of yarn used. Then they should work on solving the six-player tournament scheduling situation in Problem 2. Give them around 10 minutes to deliberate their responses.

Reconvene the class to briefly discuss the results so far. If desired this is a good time to introduce and start using the mathematical terms and definitions from graph theory, such as vertex, edge, the complete graph K_n, and edge coloring.

As the size of the tournament increases, ad hoc methods become nearly impossible to manage, and the need for an algorithmic approach quickly becomes clear. The "turning trick" algorithm is described in depth in [1]. A slightly different formulation is covered in [2]. Have students act out the turning trick by placing all but one vertex around a circle. The final vertex needs to be raised above the plane of the others to clarify this formulation. If your room has a drop ceiling, you can use it to hold the raised points; otherwise, you can have a student stand on a chair or raise his/her arm high. Each round starts by connecting a particular vertex to the final vertex, e.g., Vertex 1 in Round 1, Vertex 2 in Round 2, etc. In Round 1, the second vertex is connected to the second-to-last, the third to the third-to-last, and so on, creating parallel edges as shown in Figure 6.15. The name of the algorithm comes from the idea of turning the vertices after each round of the tournament. Therefore, Round 2 repeats the pattern of parallel edges. Figure 6.15 also illustrates the first three rounds of an eight-player tournament.

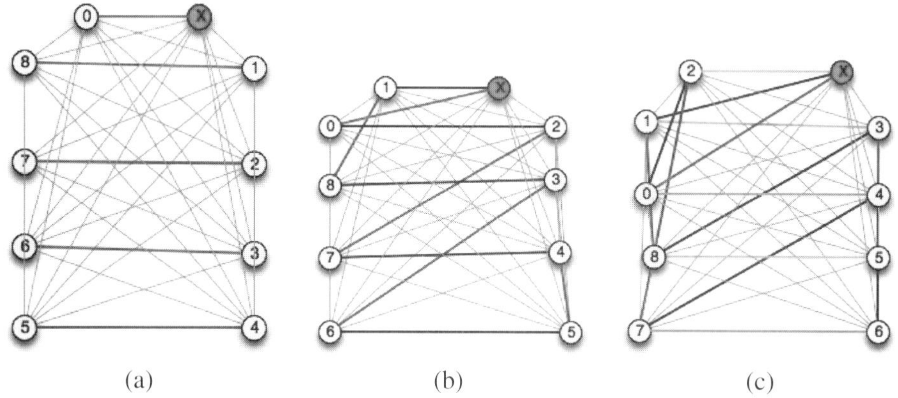

Figure 6.15: Diagram of the Turning Trick where (a) is the initial round of the tournament, (b) is the next round after one rotation, and (c) shows the pattern from continuing for one more round. Note that X remains stationary as everyone else rotates. These images are from: scienceblogs.com/goodmath/2007/07/03/edge-coloring-and-graph-turnin-1/ [3].

Now distribute Part II of the handout, and give students time to work through the two larger tournaments using the algorithm as well as tournaments that have an odd number of people. After they have solved these problems, draw out the solutions. If you are using a document camera, you can demonstrate why the algorithm was named the turning trick by drawing the first round on an overhead and spinning it. You can also demonstrate the necessity of the raised vertex by attempting to proceed without it: place all eight vertices in a circle and match them with parallel lines, turning after each round.

Suggestions and Pitfalls

Be generous in the length of the yarn! Students will want to move around and rearrange themselves.

For a small class, you may simply give the students skeins of yarn and scissors without providing specific instruction on how to visualize the tournament. This makes the activity feel more discovery-oriented.

Students in a discrete math course are often seeing algorithms for the first time. By completing this activity, students will have worked through an example of a non-trivial algorithm from problem to general solution. This is a nice visual introduction to the algorithmic nature of graph theory.

Students in a graph theory course will be exposed to many algorithms throughout the semester. This activity can fit in near the beginning of the course, serving as an introduction to graph algorithms. An early theorem to prove is the edge chromatic number of a complete graph. Given the turning trick algorithm presented in this activity, a constructive proof becomes almost trivial.

References

1. N. Hartsfield and G. Ringel, *Pearls in Graph Theory: A Comprehensive Introduction.* Dover, 2003.
2. O. Oye, *Graphs and Their Uses*, Mathematical Association of America, 1990.
3. M. C. Chu-Carroll, Edge coloring and graph turning, Science Blogs, scienceblogs.com /goodmath/2007/07/03/edge-coloring-and-graph-turnin-1/. Accessed September 24, 2015.

Designing Round-Robin Tournaments Using Yarn – Class Handout

Part I

Divide into groups to do this activity, and obtain a set of yarn and pens. Designate at least one student as a record-keeper; you don't want your results to get lost in the shuffle. Remember that a *round-robin* tournament is a competition where every pair of players or teams will eventually play one another. An individual player plays at most once per round. Use the colored pens to distinguish between rounds as you draw pictures describing each tournament.

1. As a class, we designed a four-person tournament. Draw a color-coded picture describing the tournament. How many rounds were needed? How many total games were played?

2. Expand the tournament to include six people. Fill in the table and draw a picture. How many rounds were needed? How many total games were played? What difficulties did you run into, if any?

Round	Game 1	Game 2	Game 3
Round 1			
Round 2			
Round 3			
⋮	⋮	⋮	⋮

Part II

As the size of the tournament increases, the complexity of the problem quickly becomes unmanageable without an algorithm, a set of step-by-step instructions for solving a problem.

Now you will use an algorithm known as the "turning trick" to design a tournament. Number the participants from 1 through 6. Student 6 will stand on a chair to hold the final vertex above the others. Connect participants according to this pattern:

Round 1	$1 \to 6$	$2 \to 5$	$3 \to 4$
Round 2	$2 \to 6$	$3 \to 1$	$4 \to 5$
Round 3	$3 \to 6$	$4 \to 2$	$5 \to 1$
Round 4	$4 \to 6$	$5 \to 3$	$1 \to 2$
Round 5	$5 \to 6$	$1 \to 4$	$2 \to 3$

Now that you've seen a specific instance of the algorithm, here is a general version for any tournament with an even number of participants. Number the participants from 1 through $2n$. Connect participants according to this pattern:

Round 1	$1 \to 2n$	$2 \to 2n-1$	$3 \to 2n-2$	$4 \to 2n-3$	...	$n \to n-1$
Round 2	$2 \to 2n$	$3 \to 1$	$4 \to 2n-1$	$5 \to 2n-2$	...	$n+1 \to n$
Round 3	$3 \to 2n$	$4 \to 2$	$5 \to 1$	$6 \to 2n-1$	...	$n+2 \to n+1$
$\vdots$	$\vdots$	$\vdots$	$\vdots$	$\vdots$	$\vdots$	$\vdots$

If your calculations go past $2n-1$, restart the count at 1. Why does this work?

For each of the following, make a color-coded picture of your solution.

3. Use the algorithm to design an eight-person tournament. What value should you give n?

4. Design a five-person tournament. How is this tournament different from the previous ones?

5. Design a seven-person tournament. (Hint: it may be easier than previous large tournaments!)

6.10 Constructing Disjoint Hamiltonian Cycles of Complete Graphs Using Yarn

Jill Bigley Dunham, Chapman University

Concepts Taught: graph theory, edge coloring, graph decomposition, Hamiltonian cycles, algorithmic thinking

Activity Overview

In a proof-based graph theory course, students often have difficulty keeping track of the actual problem they are trying to solve. Graph theory can be very visual, but it is still easy to get lost in the details of the definitions and algorithms.

In this activity, students create complete graphs using themselves to represent the vertices and colored yarn to represent the edges. Colors of yarn are used to decompose the graph into disjoint Hamiltonian cycles. The small cases of K_5 and K_7 can be worked out ad hoc. For larger cases and in general, the turning trick algorithm can be modified to produce disjoint Hamiltonian cycles of K_{2n+1}. This also leads us to a similar result for even complete graphs, K_{2n}. After completing this activity, students should have an intuitive understanding of the algorithm and be prepared to prove related results.

Supplies Needed (per group)	Class Time Required	Group Size
36 strands of colored yarn* 4 colored pens* 1 gallon-sized sealable bag*	50 minutes	8-10 students

*Each group will need nine two-meter long strands of each of four colors. They also need four different colored pens to match the four colors of yarn. Distribution of supplies runs most smoothly if all the materials for each group are sealed into a gallon size plastic bag prior to class.

Running the Activity

Introduce the activity by reviewing terminology and stating the two theorems we will be demonstrating and preparing to prove:

Theorem 1. The complete graph K_{2n+1} has a decomposition into n disjoint Hamiltonian cycles.

Theorem 2. The complete graph K_{2n} has a decomposition into n disjoint Hamiltonian paths and a decomposition into $n-1$ disjoint Hamiltonian cycles and a matching.

Figure 6.16: Students creating disjoint Hamiltonian cycles.

Divide the students into groups and provide them with Part I of the handout. Then have students work through the first four problems. As they work, they should begin to notice that the complexity increases as the number of vertices increases.

Once students finish the first four problems, explain how the turning trick algorithm can be applied to generate disjoint Hamiltonian cycles of the complete graph K_{2n+1}. The turning trick algorithm and proofs of these theorems are described in depth in [1].

To use the turning trick, start by placing all but one vertex around a circle. The final vertex is placed some distance away. When the students build this with yarn as shown in Figure 6.16, the final vertex is raised above the plane of the others, which makes this formulation clearer. If your room has a drop ceiling, you can use it to hold the raised points; otherwise, you can have a student stand on a chair or raise his/her arm high. As seen in the table on the handout, the jth cycle starts by connecting the raised vertex to the jth vertex, e.g., Vertex 1 in Cycle 1, Vertex 2 in Cycle 2, etc. Then use a different color to indicate the next cycle. Continue creating this graph by connecting the jth vertex to the $(2n - j)$th vertex, then continuing on to the $(j + 1)$st. The name of the algorithm comes from the idea of turning the circle of vertices after each cycle has been completed. Therefore, Cycle 2 visually repeats the pattern of edges from the first cycle.

Provide students with Part II of the handout. This includes the numerical pattern. Draw your students' attention to the two alternating patterns: the entries in the even columns are increasing while the entries in the odd columns are decreasing. Have the students work through the rest of the handout in groups.

Suggestions and Pitfalls

Be generous in the length of the yarn! Students will want to move around and rearrange themselves.

If there is time after they have solved these problems, reconvene the class to discuss and draw the solutions. If you are using a document camera, you can demonstrate why the algorithm was named the turning trick by drawing the first round on an overhead and spinning it.

Once they have solved these problems, students are prepared to prove Theorems 1 and 2, along with related results.

Reference

1. N. Hartsfield and G. Ringel, *Pearls in Graph Theory: A Comprehensive Introduction.* Dover, 2003.

Constructing Disjoint Hamiltonian Cycles of K_{2n+1} – Class Handout

Part I

Obtain a set of yarn and pens for your group. Designate one group member as the record-keeper.

Theorem 1. The complete graph K_{2n+1} has a decomposition into n disjoint Hamiltonian cycles.

1. Start by creating a graph with five (human) vertices. How many disjoint Hamiltonian cycles should you be able to find? Create them with different colors of yarn. Draw a picture of your result using the colored pens.

2. Add two human vertices to your graph and create the decomposition as above. See if you can extend your previous configuration instead of starting from scratch (if not, starting from scratch is valid as well!). Draw a picture of this result using the colored pens.

3. Theorem 1 says nothing about the decomposition of a complete graph with an even number of vertices. What can you say about the degree of each vertex in a complete graph? What does this say about the decomposition of an even complete graph K_{2n} into disjoint Hamiltonian cycles?

Theorem 2. The complete graph K_{2n} has a decomposition into n disjoint Hamiltonian paths and a decomposition into $n-1$ disjoint Hamiltonian cycles and a matching.

4. Remove one vertex from your decomposed K_7. Draw a picture of the configuration using the colored pens. Which decomposition does this K_6 demonstrate?

Part II

As the size increases, the complexity quickly becomes unmanageable without employing an algorithm.

Algorithm for decomposing K_{2n+1} into disjoint Hamiltonian cycles:
Number the participants from 1 through $2n+1$. Student $2n+1$ will stand on a chair to hold the final vertex above the others. Connect participants according to the following pattern, closing each cycle:

Cycle 1	$2n+1$	1	$2n$	2	$2n-1$	...	$n+1$
Cycle 2	$2n+1$	2	1	3	$2n$	...	$n+2$
Cycle 3	$2n+1$	3	2	4	1	...	$n+3$
$\vdots$	$2n+1$	$\vdots$	$\vdots$	$\vdots$	$\vdots$	$\vdots$	$\vdots$
Cycle n	$2n+1$	n	$n-2$	$n+1$	$n-3$	...	$2n$

If your calculations go past $2n$, wrap back around to count from 1 again.

5. Use the algorithm to re-create a decomposed K_7. Is it equivalent to your previous decomposition up to isomorphism?

6. Remove one vertex from your decomposed K_7. *Hint: There may be an obvious candidate here.* Draw a picture of the configuration using the colored pens. Which decomposition does this K_6 demonstrate?

7. Use the algorithm to create a decomposed K_9. Draw a picture of the configuration using the colored pens.

8. Remove one vertex from your decomposed K_9. Which decomposition does this K_8 demonstrate?

9. Using this algorithm to decompose a K_{2n} should always result in the same decomposition. Which one?

6.11 Exploring the ϵ-N Definition of Sequence Convergence with Yarn

Julie Barnes, Western Carolina University

Concepts Taught: sequences, definition of sequence convergence

Activity Overview

Although senior level mathematics students have seen sequences at some point in a calculus class, they typically do not have a lot of experience with them. Yet, in a real analysis class, they are asked to jump into a formal definition of sequence convergence, and that can be intimidating for them. In this activity, students have a chance to become the first few elements of a sequence and use yarn to demonstrate the formal definition of convergence. Not only does this emphasize the discrete nature of sequences, but it also helps students see the dynamic relationship between ϵ and N in the definition.

Supplies Needed	Class Time Required	Group Size
18 feet of adding machine paper Tape Prepared ϵ region*	15-20 minutes	Class demonstration with 6-8 volunteers

*The ϵ region needs to be constructed before class. Start by cutting two pieces of yarn, each roughly 10 feet long. Tie a small object to the end of each piece of yarn as seen in Figure 6.17a. This prevents the yarn from knotting up in storage. Door knob hangers from a craft aisle work very well because the yarn can be wrapped around it without creating knots; wooden letters or even a simple piece of cardboard work as well.

Before class, find a location approximately $10' \times 8'$ where you can create a coordinate system on the floor. Tape the adding machine paper on the floor to mark off a $10'$ x-axis and an $8'$ y-axis. Construct the coordinate system so that the x-axis is almost all positive. The y-axis should at least have 5 feet in the positive region. Mark units roughly 1 foot apart on each axis with a marker. See Figure 6.17b for an example of this setup. Painter's tape can be used instead of adding machine paper.

Running the Activity

Provide students with copies of the handout that contains the examples used in the class demonstration. For each example, have 4-6 volunteers become the first few terms of a sequence of points by having one student stand at the location of the first point of the sequence, a second student stand at the location of the second point, and so on. In addition, two student volunteers will act as the *epsilon pair*. The epsilon pair will be holding the ϵ region that you prepared in advance as shown in Figures 6.17b, 6.17c, and 6.17d. The rest of the class should stand in a location where they can see all the volunteers. As a class, have students discuss the questions posed on the handout while they explore what happens to N as the ϵ region changes.

Suggestions and Pitfalls

Students typically need assistance in determining where they should stand. It helps to ask the class as a whole what it means to be a sequence. Many have not made the connection that a sequence is a function on the natural numbers, and they are surprised that they can simply evaluate their position number and represent the term by standing at the location described as (n, x_n).

Remind students that a sequence is not finite and that they are only representing the first few terms of the sequence in this activity. These representations should give students a feel for the long term behavior of these sequences, but they are not the entire sequences.

For a more information about this activity, as well as an extension using sticky notes, see [1] where this activity first appeared.

Reference

1. J. Barnes, Feather boas in real analysis, *PRIMUS* 21 no. 2 (2011) 130–141.

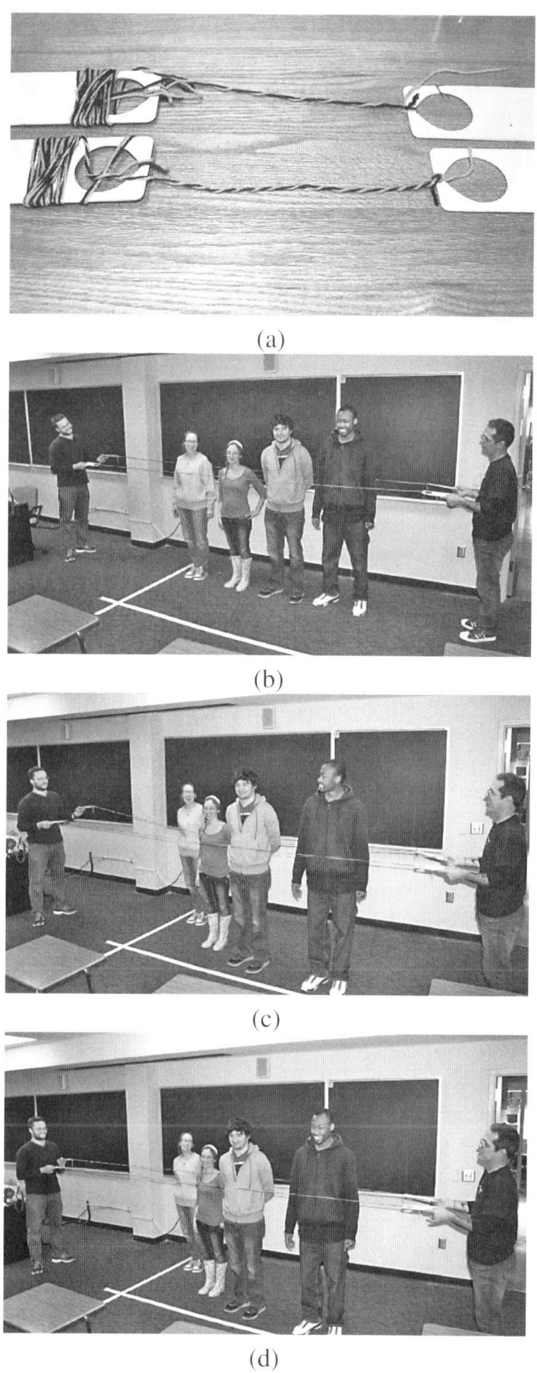

(a)

(b)

(c)

(d)

Figure 6.17: (a) Setup for ϵ region. (b) Students demonstrating that for a constant sequence, all points fall in the ϵ region no matter what the value of ϵ is. (c) Students finding N for a given large ϵ. Note that three students are between the yarn. (d) Students finding N for a given smaller ϵ. Now only two students are within the yarn.

Exploring the $\epsilon - N$ Definition of Sequence Convergence – Class Handout

1. Let $a_n = 3$. As a class, we will represent this sequence by having one student stand at each of the points $(1,3)$, $(2,3)$, $(3,3)$, $\ldots$. Two people (referred to as the *epsilon pair*) will use string to represent a large ϵ region. Since this sequence converges to 3, the epsilon pair needs to stand on the edges of the coordinate system along the line $y = 3$. They should start with their arms wide open, representing a large ϵ region.

   (a) Are all the students from the sequence inside the ϵ region?

   (b) What happens as the students representing the ϵ region make the region smaller? Are all the students from the sequence still in the ϵ region?

   (c) Is it possible to find an ϵ region small enough so that someone from the sequence will not be in the ϵ region? Why or why not?

2. Let $b_n = \frac{3}{n}$. As a class, we will represent this sequence by having one student stand at each of the points $(1, \frac{3}{1})$, $(2, \frac{3}{2})$, $(3, \frac{3}{3})$, $(4, \frac{3}{4})$, $\ldots$. Two people will use string to represent a large ϵ region. Since this sequence converges to 0, the epsilon pair needs to stand on the edges of the coordinate system along the line $y = 0$. They should start with their arms wide open, representing a large ϵ region.

   (a) Are all the students from the sequence inside the ϵ region?

   (b) What happens as the students representing the ϵ region make the region smaller? Are all the students from the sequence still in the ϵ region?

   (c) For smaller ϵ values, which students remain inside the ϵ region? What does this have to do with N?

3. Let $c_n(x) = (-1)^n$. As a class, we will represent this sequence by having one student stand at each of the points $(1,-1)$, $(2,1)$, $(3,-1)$, $(4,1)$, $\ldots$. Two people will use string to represent a large ϵ region. You may want to try 0, 1, or -1 as possible values for the limit.

   (a) Are all the students from the sequence inside the ϵ region?

   (b) Make ϵ smaller. What happens to the number of students inside the ϵ region as ϵ decreases?

   (c) Does this situation satisfy the definition of convergence? Why or why not? Does c_n converge to zero? Does c_n converge to 1? Does c_n converge to -1?

6.12 Exploring the ϵ-δ Definition of Continuity

Julie Barnes, Western Carolina University

Concepts Taught: definition of continuity

Activity Overview

Even though real analysis students are very familiar with a wide variety of continuous functions, they often find the abstract nature of the formal definition of continuity to be confusing. Drawing diagrams to explain the definition is helpful, but an image may be too static when students first see the definition. This can be addressed by using technology that makes it possible to zoom in at different points of the graph. However, the zoomed-in graph does not look like the initial function anymore, and it is possible to forget that this local region is still a portion of a larger function. In this activity, students physically act out the formal definition of continuity by creating epsilon and delta regions at a point while simultaneously standing on a large graph of the function. This gives students a way to see the dynamic relationships between epsilon and delta while preparing them to better understand the meaning of the definition.

Supplies Needed	Class Time Required	Group Size
Two feather boas* Tape 18 feet of adding machine paper** Prepared ϵ and δ regions***	15-20 minutes	Class demonstration with 4 volunteers

*Feather boas can be purchased at most craft stores. They work nicely because they stay in place fairly well and are large enough to see for a classroom demonstration. However, yarn can be used instead of the boas.
**Painter's tape can be used in place of adding machine paper.
***The ϵ and δ regions need to be constructed before class. To do this, cut four pieces of yarn that are each 10 feet long. Then tie each end to a weighted object, such as a door knob hanger, to prevent the yarn from developing knots. See Figure 6.18a for a photograph of these yarn regions. You are creating two pairs: one pair for the ϵ region and one for the δ region. It is best if one color of yarn is used for ϵ region, and another color for the δ regions.

Prior to class, find a location to do the activity that includes room for a 10 foot by 8 foot xy-coordinate system as well as space for students to stand around the coordinate system. This may involve moving some desks or going to a different location. Tape 10 feet of adding machine paper on the floor to create an x-axis and another 8 feet of adding machine paper for the y-axis. See Figures 6.18b, 6.18c, and 6.18d for the setup as well as illustrations of each step in this activity.

Running the Activity

Provide students with the handout, and use it as an outline for the activity. Ask students to record their observations on the handout. For the first part, tie two feather boas together and use them to create a function on the floor. Have two students volunteer to be the epsilon pair, and two more students be the delta pair. The epsilon pair should make an ϵ region around the knot obtained from tying the feather boas together. Make sure this first ϵ region is at least a foot wide. The delta pair then creates a δ region that is small enough to satisfy the definition of continuity. Have students work together to complete Problem 1 from the handout about this setup. For Problem 2, the goal is for students to observe dynamically how ϵ and δ are related when the function is continuous. Have the epsilon pair decrease the size of ϵ. The delta pair then needs to decrease the size of the δ region so that the definition of continuity will still be satisfied. Use the problems from the handout to guide discussion. See Figures 6.18b and 6.18c for an example of students working with a continuous function. Finally, in Problem 3 students explore what happens at a point where the function is not continuous. Untie the feather boas and create a new function that has a jump discontinuity. The epsilon pair should create an ϵ region around one edge of the jump at the point of discontinuity. Have the students try varying the size of ϵ and seeing if it is possible for the delta pair to create a region small enough to satisfy the definition of continuity. See Figure 6.18d for an example.

For a more information about this activity, see [1] where this activity first appeared.

Reference

1. J. Barnes, Feather boas in real analysis, *PRIMUS* 21 no. 2 (2011) 130–141.

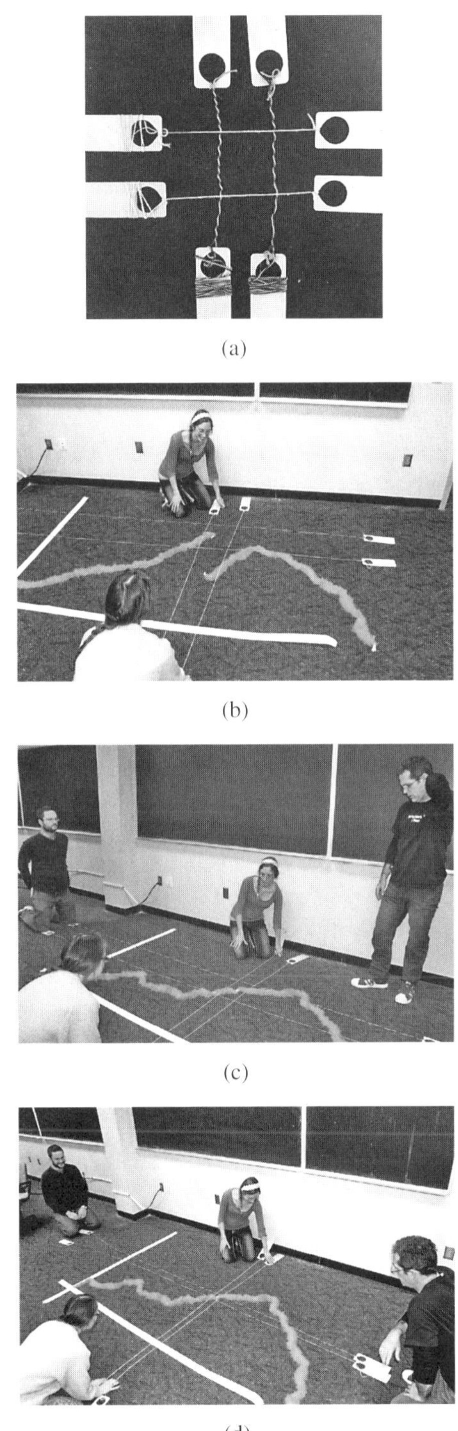

(a)

(b)

(c)

(d)

Figure 6.18: (a) Prepared ϵ and δ regions. (b) Attempting to find ϵ and δ regions for a discontinuous function. (c) Students finding δ for a fairly large ϵ. (d) Students finding δ for a smaller ϵ.

Explorations of the $\epsilon - \delta$ Definition of Continuity – Class Handout

1. A few of your classmates will create an ϵ region and corresponding δ region around a point on a continuous function that is made with feather boas on the floor. Complete the following problems concerning the setup.

   (a) Sketch the ϵ and δ regions, including the portion of the function bounded by these regions.

   (b) Describe where the function crosses the ϵ and δ regions in your diagram from Problem 1a. Does the function cross the yarn representing ϵ or the yarn representing δ or both?

   (c) What must be true about the δ region in order for the definition of continuity to hold? How can you tell that the δ region is small enough?

2. Two volunteers will decrease the size of the ϵ region. Two other volunteers then need to decrease the size of the δ region so that the definition of continuity will still be satisfied. Complete the following problems concerning what you are observing while this happens.

   (a) Sketch the ϵ and δ regions, including the portion of the function bounded by these regions.

   (b) Describe where the function crosses the ϵ and δ regions in your diagram from Problem 2a. Does the function cross the yarn representing ϵ or the yarn representing δ or both? How does your diagram in Problem 1a compare to the region you obtained in Problem 2a?

   (c) Based on the activity so far, would you say that the definition of continuity is more concerned about properties near a point or globally? How can you tell?

3. The feather boas are now used to represent a function with a jump discontinuity. Two volunteers will create a variety of different sized ϵ regions while two more volunteers attempt to find a δ region that satisfies the definition of continuity. Complete the following problems concerning what you are observing.

   (a) For very small ϵ values, is it possible to find a δ small enough to satisfy the definition of continuity? Should there be a small enough δ to make the definition work?

   (b) Is it possible to create an ϵ region that is large enough so that there is a δ region that meets the criterion of the definition of continuity? If so, does that cause a problem with the definition of continuity? Why or why not?

6.13 Walking Complex Functions

Julie Barnes, Western Carolina University
Beth Schaubroeck, United States Air Force Academy

Concepts Taught: graphing complex functions

Activity Overview

Many students find that basic algebraic computations on the complex plane are similar to those studied on the reals; therefore performing algebraic computations do not tend to cause students much difficulty. However, the geometric meaning behind these computations does not follow the same patterns as seen in functions on the reals. In this activity, students who already are familiar with algebraic computations use a kinesthetic approach to better visualize the geometric behavior of complex functions.

Each student begins at a point on the complex plane, evaluates a given function at his/her point, and then moves to the location representing the function's output. Students literally feel how their points move from one location to another, while simultaneously observing how other points are moving. This activity is designed to help students see beyond the algebra and view functions more dynamically.

Supplies Needed	Class Time Required	Group Size
40 feet of adding machine paper Tape	10-15 minutes	Whole class

This activity requires enough floor space for students to stand about an arm's length from each other. This may be accomplished by moving desks to one side of the room or relocating students to a hallway or other location. Before class, designate where the real and imaginary axes are located. Then tape adding machine paper on the floor to create real and imaginary axes. Painter's tape can be used instead of adding machine paper. Mark units on the axes roughly one foot apart with a marker.

Running the Activity

Provide each student with a copy of the handout, which has a list of functions that the class will be representing. Have students spread out in the prepared complex plane. Ensure that one student is at the origin, at least one is on the positive real axis, and at least one is on the positive imaginary axis. Everyone else should spread out, with some near the origin and some far away. Have each student record the coordinate of their initial location on the handout.

Begin with the first function, $f_1(z) = z$, and have each student compute the output of $f_1(z)$ at his/her point. Since the first function is the identity, nobody's point will change value. Ask students to move to their new value, and nobody should move. Then, give students a chance to record any observations they made about $f_1(z)$.

Change the function to $f_2(z) = z + 1$. This time points will be affected by the function. Give everyone time to evaluate $f_2(z)$ at his/her point and to help each other if necessary. Once everyone is ready, move together as a group. Make observations as a whole and discuss what happened. Give students a chance to record any general observations. Once it is clear to everyone that this function moves all points in the positive real direction by one unit, have everyone go back to their initial location.

Continue working through the functions listed on the handout. After the students have moved to $f(z)$ for each function, stop and ask questions about what happened. Did everyone move in the same direction? Did everyone move away from the origin? Did everyone stay the same distance from the origin? Did the points rotate? Did some people move away from the origin while others moved closer to the origin? Give students a chance to take notes. Figure 6.19 shows an example of students' final destinations after applying some of the functions listed in the handout.

Suggestions and Pitfalls

If your class has more than 30 students, this activity could be done as a demonstration with a 6-10 volunteers.

Students must mentally process enacting a function differently than they would in graphing by hand. For example, when students walk $z + 1$, they are often still thinking of the function $g(x) = x + 1$ from calculus, and they may try to walk parallel to the imaginary axis instead of parallel to the real axis. When they enact iz, students are often impressed to see everyone rotate together; this emphasizes the fact that multiplying by i is a rotation. For $2z$, students often have to think a little about what it means; they may find it interesting that all points move radially away from the origin. For $\text{Im}(z)$, students tend to walk to the imaginary axis instead of the real axis; they have trouble thinking of the imaginary part being a real number instead of a purely imaginary number. Students find $\frac{1}{z}$ to be particularly difficult and may need more assistance.

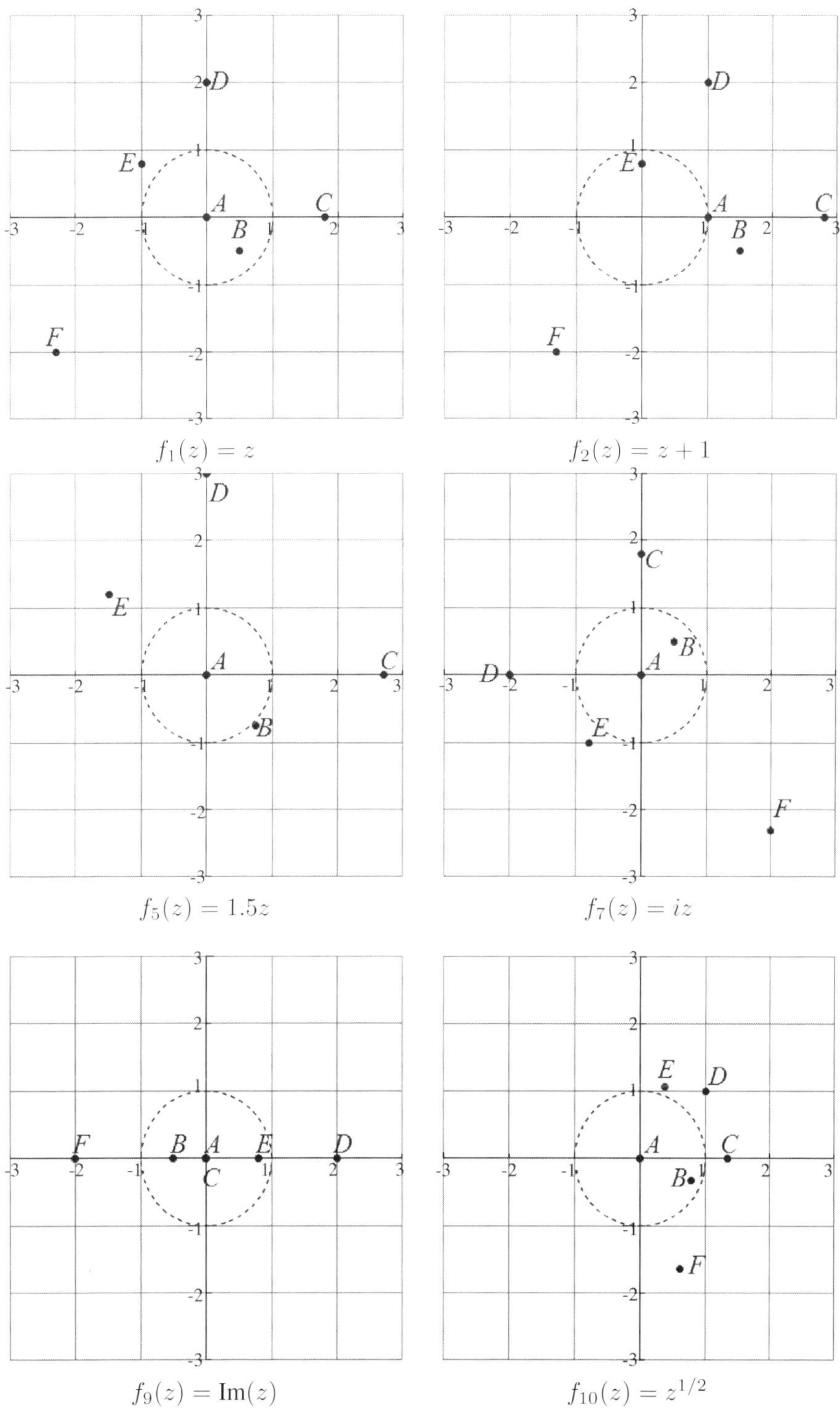

Figure 6.19: Examples of where given points move under some complex functions from the handout. The function names in this diagram match the names on the handout.

Walking Complex Functions – Handout

My starting point for this activity is ________________________________

Use the blank space to take notes about what happens.

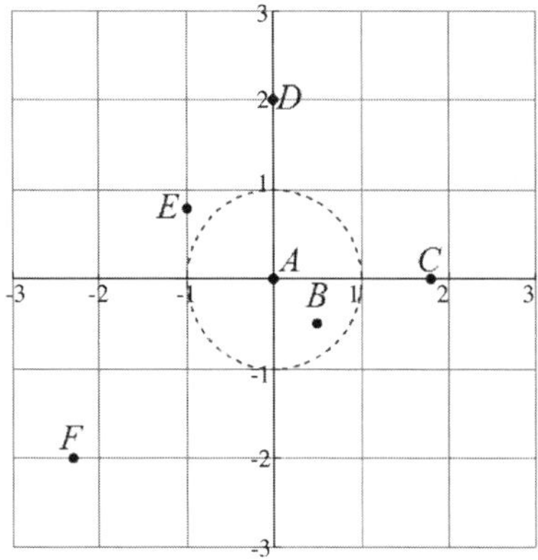

1. $f_1(z) = z$

2. $f_2(z) = z + 1$

3. $f_3(z) = z + i$

4. $f_4(z) = \bar{z}$

5. $f_5(z) = 1.5z$

6. $f_6(z) = \frac{1}{2}z$

7. $f_7(z) = iz$

8. $f_8(z) = \text{Re}(z)$

9. $f_9(z) = \text{Im}(z)$

10. $f_{10}(z) = z^{1/2}$

11. $f_{11}(z) = z^2$

12. $f_{12}(z) = \frac{1}{z}$

6.14 Graphing Complex Functions with Feather Boas

Julie Barnes, Western Carolina University
Beth Schaubroeck, United States Air Force Academy

Concepts Taught: graphing complex functions

Activity Overview

Visualizing complex functions is challenging because the graph of a complex function is four dimensional. One visualization technique is to sketch pictures of both the domain and range, showing how a set of points in the domain is transformed by the function to a new image in the range. In this activity, students consider a set S as the domain and then plot the image of that set under a variety of common complex functions by using feather boas or other objects to represent the range. This activity is designed to help students develop a more concrete understanding of complex functions. After students have completed this activity, it is useful to provide them with technological tools that will produce the images of sets more precisely.

Supplies Needed (per group)	Class Time Required	Group Size
2 feather boas 4 pieces of tape 10 feet of adding machine paper, cut into to 5″ long pieces 2 strips of paper, each roughly $1'' \times 6''$ 4 blocks	40-50 minutes	2-4 students

This activity can be done with a wide variety of objects, not only the ones listed above or shown in the photographs. Feather boas work well because they are pliable, keep their shape, and students enjoy working with them. Feather boas can be purchased at most craft stores. Some alternatives are yarn, clothesline, or holiday garlands. Blocks can be replaced by anything fairly small that won't roll away, such as coins. Also, painter's tape can be used instead of adding machine paper.

Running the Activity

Provide each group with four pieces of tape, two five-foot long pieces of adding machine paper, two feather boas, two strips of paper, four blocks, and the handout. Have the students tape the adding machine paper to the floor to create the real and imaginary axes. Once the axes are ready, students will use the materials provided to create representations of the image of a given set under each of the functions indicated on the handout. The students should be concerned about the basic shape and general trends; this is not precise work. While the students work, circulate around the room to answer questions and check their progress, offering hints or asking guiding questions when you see mistakes or encounter groups that are stuck. See Figure 6.20 for some solutions to the handout using feather boas, paper strips, and blocks.

Suggestions and Pitfalls

Space is always an issue. If your classroom is large enough with movable desks, have the students move the desks aside and use the floor. Depending on constraints at your institution, students may be able to move into the hallway or other nearby spaces such as conference rooms or study areas. Alternatively, if the room has large tables, students can use those instead of the floor.

There will always be physical constraints when dealing with objects like feather boas and strips of paper. For example, students may have trouble stretching a feather boa to represent z^2, or shrinking a boa to represent $z^{1/2}$. This provides a good starting point for a discussion of how the functions really behave.

A common student misconception occurs with $f_5(z) = \mathrm{Im}(z)$. Students often depict the output as lying on the imaginary axis instead of the real axis. Be ready to ask students to think about what $\mathrm{Im}(z)$ means.

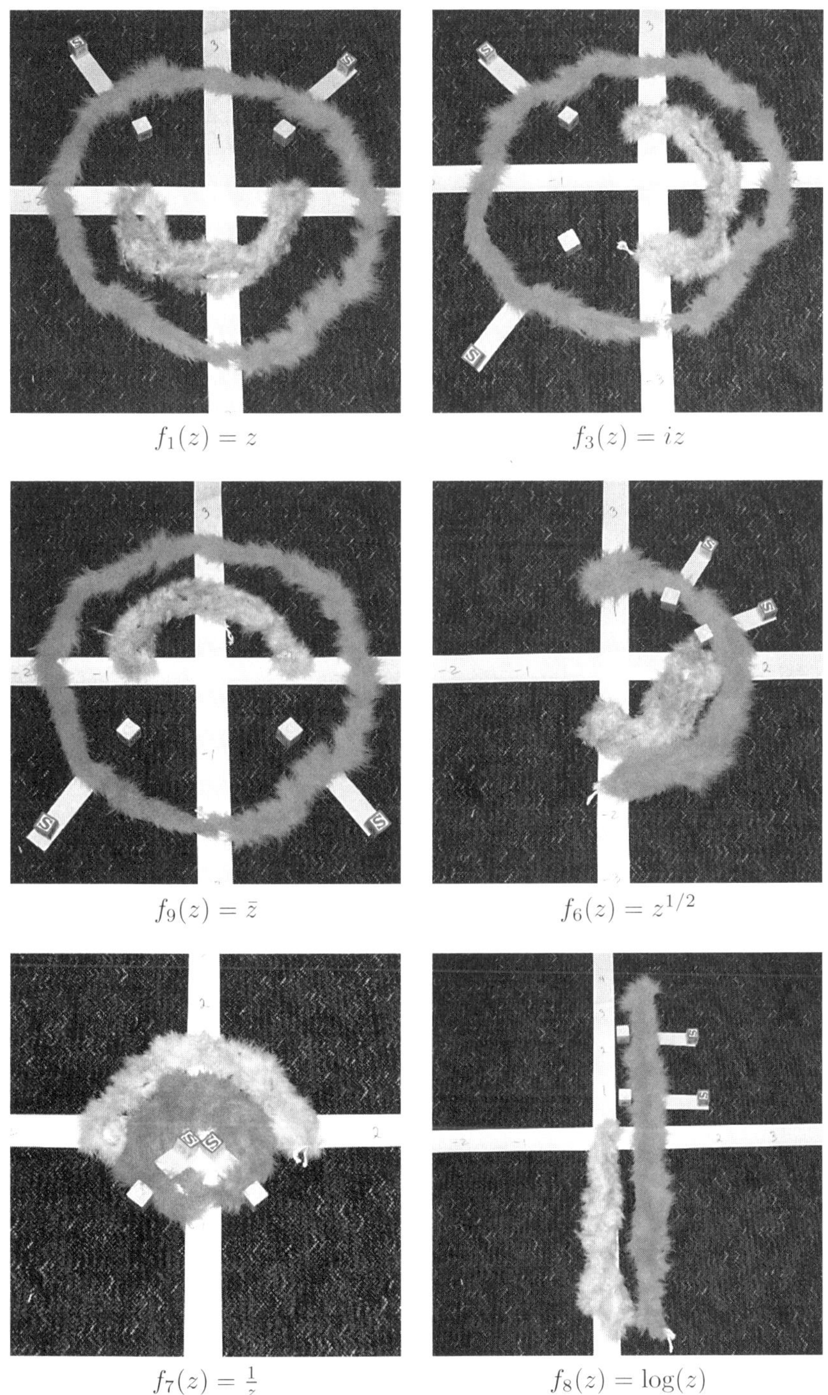

Figure 6.20: Some examples of graphs of complex functions created with feather boas. The graphs are not exact, but they show the essence of what is occurring with these functions. The function names in the figure correspond with the function names in the handout.

Graphing Complex Functions with Feather Boas – Class Handout

Use two pieces of adding machine paper that are about five feet long to create the real and imaginary axes. Tape them in place. Let S be the set of points in the set provided below. For each function below, use the materials provided to represent the image of the set S under the given function. After your team completes each graph with the materials provided, make a sketch of the resulting image.

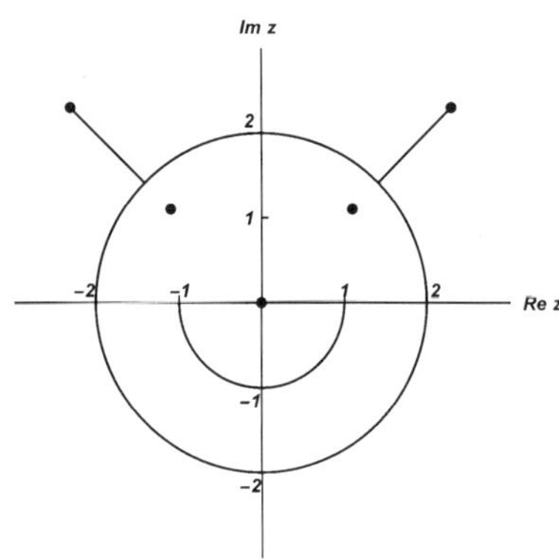

1. $f_1(z) = z$

2. $f_2(z) = z^2$

3. $f_3(z) = iz$

4. $f_4(z) = iz + 1$

5. $f_5(z) = \mathrm{Im}(z)$

6. $f_6(z) = z^{1/2}$

7. $f_7(z) = \frac{1}{z}$

8. $f_8(z) = \log(z)$

9. $f_9(z) = \bar{z}$

10. Keeping in mind that the image in the domain is symmetrical, find another function, $f_{10}(z)$, that transforms the given domain to the same range seen in Problem 9. Would this new function send *all* domains to the same image as, $f_9(z) = \bar{z}$ does?

6.15 Exploring Knots

Hannah Robbins, Roanoke College

Concepts Taught: knot theory, invariants

Activity Overview

In this activity, students experiment with a collection of physical knots to explore the basic ideas of knot theory, especially what it means for two knots to be the same or different. This allows them to experience the process of exploring a completely new mathematical subject via experimentation and examples. In an upper level course such as topology, this activity can also be used to introduce the concept of mathematical invariants.

Supplies Needed (per group)	Class Time Required	Group Size
One bag of knots*	30 minutes	2 students

*Before this activity begins, use clothesline, scissors, athletic tape, and markers to create color coded knots for each group; resealable clear gallon sized bags work well for storing and distributing the knots. Each bag should have one standard unknot, one figure-eight knot, two trefoil knots color coded with different colors, one Hopf link, one Solomon's "knot" which is actually a link, two Whitehead links color coded with different colors, and one Borromean rings link. Also include one "knotted" unknot, which is a standard unknot tangled up. These knots are shown in Figure 6.21a. Use the markers to color code the knots by type (Figure 6.21b); it is very helpful to make a key matching the colors of the knots to their type for future reference.

Making the bags of knots when you first do this activity can take about 10-15 minutes per bag, but the bags can be reused many times. Just remember to tangle up the "knotted" unknots again.

Running the Activity

As a class, briefly introduce that a mathematical knot is a piece of string with the ends connected. Explain that they are allowed to manipulate their knots in any way they like except for cutting the string or disconnecting the ends.

Have the students split up into pairs, and give each pair one knot bag and a copy of the handout. Tell them it is their job to figure out which knots are "the same" and which "are different". Give them time to work on this problem in their pairs. If any pairs seem stuck you can ask them:

(a)

(b)

Figure 6.21: Types of knots.

Are there any knots that you can completely untangle?

Are there any knots that you know are the same? How can you tell?

Are there any knots that you are sure cannot be the same? How can you tell?

After the pairs have reached some conclusions, reconvene as a class. Discuss the definitions of "the same" and "different" each pair used. Have different pairs volunteer knots they think are the same or different and explain how they can tell. This is where the color coding is vital, because everyone can refer to the knots by color!

Students usually recognize the three pairs of knots {unknots, trefoils, Whitehead links} and explain that they are the same because they can be rearranged to look identical. They also usually realize that the number of complete loops of string in a knot cannot be changed by rearranging it. This lets them distinguish between any knot with one loop {unknots, figure-eight, trefoils} and any knot with two loops {Hopf link, Solomon's knot, Whitehead links}. It also proves that the Borromean rings link is unique, since it is the only one with three loops.

Suggestions and Pitfalls

Groups of two work best because both people can easily get their hands on the knots. Groups of three are okay, but not as good. I wouldn't suggest groups of four or more.

Be stingy with specifics! Don't define what makes two knots "the same" or "different". Make it clear to the students that part of their job is to develop these definitions. It is great to point out that demonstrating how two knots can be rearranged to look identical proves they are the same, but claiming you can't figure out how to make them look identical doesn't prove they are different. This idea can be tied to more general proofs if your class's level permits.

This activity can easily be adapted to classes with different levels of mathematical proficiency. In an advanced class such as topology or geometry, try including formal vocabulary like "invariant" and "homeomorphism". In other settings such as a liberal arts math course, you can discuss the same ideas with everyday language like "untangling". If you are using this activity as a lead-in to a more detailed investigation of knot theory, you can make one of your trefoils "left" and the other "right". Then you can also discuss their more nuanced relationship since now they are not the same knot despite their clear relationship. By contrast, a figure-eight knot's "left" and "right" versions are the same.

Exploring Knots – Class Handout

Your group has a bag containing a collection of knots, each coded with a different color. The colors allow you to refer to a specific knot when discussing things with your group or the class.

In mathematics, a knot is defined as a piece of string with the ends connected. You can do anything you like to your knots except you are not allowed to cut the string or disconnect the ends. Your job is to figure out which knots are the same and which are different. To get started, try to answer the questions below. Remember that you can refer to the knots by color.

1. Are there any pairs of knots that you think are the same? Which ones? How can you tell?
2. What does it mean for two knots to be "the same"?
3. Are there any pairs of knots that you think are "different"? Which ones? How can you tell?
4. What does it mean for two knots to be "different"?

Concept Index

Author Index

Main Ingredient Index

Course Index